W9-AFI-368

FOUNDATIONS OF THREE-DIMENSIONAL EUCLIDEAN GEOMETRY

MONOGRAPHS AND TEXTBOOKS IN PURE AND APPLIED MATHEMATICS

1. *K. Yano,* Integral Formulas in Riemannian Geometry (1970)
2. *S. Kobayashi,* Hyperbolic Manifolds and Holomorphic Mappings (1970)
3. *V. S. Vladimirov,* Equations of Mathematical Physics (A. Jeffrey, editor; A. Littlewood, translator) (1970)
4. *B. N. Pshenichnyi,* Necessary Conditions for an Extremum (L. Neustadt, translation editor; K. Makowski, translator) (1971)
5. *L. Narici, E. Beckenstein, and G. Bachman,* Functional Analysis and Valuation Theory (1971)
6. *D. S. Passman,* Infinite Group Rings (1971)
7. *L. Dornhoff,* Group Representation Theory (in two parts). Part A: Ordinary Representation Theory. Part B: Modular Representation Theory (1971, 1972)
8. *W. Boothby and G. L. Weiss (eds.),* Symmetric Spaces: Short Courses Presented at Washington University (1972)
9. *Y. Matsushima,* Differentiable Manifolds (E. T. Kobayashi, translator) (1972)
10. *L. E. Ward, Jr.,* Topology: An Outline for a First Course (1972) *(out of print)*
11. *A. Babakhanian,* Cohomological Methods in Group Theory (1972)
12. *R. Gilmer,* Multiplicative Ideal Theory (1972)
13. *J. Yeh,* Stochastic Processes and the Wiener Integral (1973) *(out of print)*
14. *J. Barros-Neto,* Introduction to the Theory of Distributions (1973) *(out of print)*
15. *R. Larsen,* Functional Analysis: An Introduction (1973)
16. *K. Yano and S. Ishihara,* Tangent and Cotangent Bundles: Differential Geometry (1973)
17. *C. Procesi,* Rings with Polynomial Identities (1973)
18. *R. Hermann,* Geometry, Physics, and Systems (1973)
19. *N. R. Wallach,* Harmonic Analysis on Homogeneous Spaces (1973)
20. *J. Dieudonné,* Introduction to the Theory of Formal Groups (1973)
21. *I. Vaisman,* Cohomology and Differential Forms (1973)
22. *B.-Y. Chen,* Geometry of Submanifolds (1973)
23. *M. Marcus,* Finite Dimensional Multilinear Algebra (in two parts) (1973, 1975)
24. *R. Larsen,* Banach Algebras: An Introduction (1973)
25. *R. O. Kujala and A. L. Vitter (eds.),* Value Distribution Theory: Part A; Part B. Deficit and Bezout Estimates by Wilhelm Stoll (1973)
26. *K. B. Stolarsky,* Algebraic Numbers and Diophantine Approximation (1974)
27. *A. R. Magid,* The Separable Galois Theory of Commutative Rings (1974)
28. *B. R. McDonald,* Finite Rings with Identity (1974)
29. *J. Satake,* Linear Algebra (S. Koh, T. Akiba, and S. Ihara, translators) (1975)

30. *J. S. Golan,* Localization of Noncommutative Rings (1975)
31. *G. Klambauer,* Mathematical Analysis (1975)
32. *M. K. Agoston,* Algebraic Topology: A First Course (1976)
33. *K. R. Goodearl,* Ring Theory: Nonsingular Rings and Modules (1976)
34. *L. E. Mansfield,* Linear Algebra with Geometric Applications (1976)
35. *N. J. Pullman,* Matrix Theory and Its Applications: Selected Topics (1976)
36. *B. R. McDonald,* Geometric Algebra Over Local Rings (1976)
37. *C. W. Groetsch,* Generalized Inverses of Linear Operators: Representation and Approximation (1977)
38. *J. E. Kuczkowski and J. L. Gersting,* Abstract Algebra: A First Look (1977)
39. *C. O. Christenson and W. L. Voxman,* Aspects of Topology (1977)
40. *M. Nagata,* Field Theory (1977)
41. *R. L. Long,* Algebraic Number Theory (1977)
42. *W. F. Pfeffer,* Integrals and Measures (1977)
43. *R. L. Wheeden and A. Zygmund,* Measure and Integral: An Introduction to Real Analysis (1977)
44. *J. H. Curtiss,* Introduction to Functions of a Complex Variable (1978)
45. *K. Hrbacek and T. Jech,* Introduction to Set Theory (1978)
46. *W. S. Massey,* Homology and Cohomology Theory (1978)
47. *M. Marcus,* Introduction to Modern Algebra (1978)
48. *E. C. Young,* Vector and Tensor Analysis (1978)
49. *S. B. Nadler, Jr.,* Hyperspaces of Sets (1978)
50. *S. K. Sehgal,* Topics in Group Rings (1978)
51. *A. C. M. van Rooij,* Non-Archimedean Functional Analysis (1978)
52. *L. Corwin and R. Szczarba,* Calculus in Vector Spaces (1979)
53. *C. Sadosky,* Interpolation of Operators and Singular Integrals: An Introduction to Harmonic Analysis (1979)
54. *J. Cronin,* Differential Equations: Introduction and Quantitative Theory (1980)
55. *C. W. Groetsch,* Elements of Applicable Functional Analysis (1980)
56. *I. Vaisman,* Foundations of Three-Dimensional Euclidean Geometry (1980)

Other Volumes in Preparation

FOUNDATIONS OF THREE-DIMENSIONAL EUCLIDEAN GEOMETRY

Izu Vaisman

Department of Mathematics
University of Haifa
Haifa, Israel

MARCEL DEKKER, INC. New York and Basel

Library of Congress Cataloging in Publication Data

Vaisman, Izu.
Foundations of three-dimensional Euclidean geometry.

(Pure and applied mathematics ; 56)
Bibliography: p.
Includes index.
I. Geometry--Foundations. I. Title.
QA681.V3 516.2 80-18890
ISBN 0-8247-6901-0

MARCEL DEKKER, INC.

270 Madison Avenue, New York, New York 10016

Current printing (last digit):

10 9 8 7 6 5 4 3 2 1

PRINTED IN THE UNITED STATES OF AMERICA

PREFACE

Three-dimensional Euclidean Geometry is one of the oldest chapters of mathematics. Not only its beginning, but also its first systematic development was achieved in the ancient era by Euclid's celebrated *Elements* of the third century B.C. However, the rigorous logical foundations of this geometry were established only at the end of the 19th century by the classical work of David Hilbert *Grundlagen der Geometrie* (1899) [9].

The present book aims to present to the reader a modern axiomatic construction of three-dimensional Euclidean geometry, in a rigorous and accessible form. Its leading principle consists of the gradual formulation of the axioms such that, by determining step by step all the geometric spaces satisfying the considered groups of axioms, we can finally characterize, up to an isomorphism, the Euclidean space.

In carrying out this program, our main sources were the above mentioned book by D. Hilbert [9], and the books by E. Artin [1], and C. Teleman [16, Chapter III]. As for the geometric part of the book, we are most indebted to B. Kerékjártó [10].

The mentioned theory is sometimes presented here in a nonclassical manner, and new elements appear in the chapters devoted to order axioms, to congruence axioms, etc. The reader will find in the first paragraph of the Introduction a more precise description of the place of our book within the flow of ideas and theories whose ultimate result is the modern construction of the foundation of Euclidean geometry.

Like in all the books treating the same subject, we use drawings, but they have a purely didactic, auxiliary character, while the reasoning is based only on the axioms accepted. We employ the usual system of referring to the results obtained. For instance, Theorem 1.1.1 means the Theorem 1 of §1 of the Chapter 1. Within the same chapter we do not indicate the chapter, and within the same section we also do not indicate the section.

Exercises and problems are added to each chapter, and the reader is also invited to give by himself or herself the proofs of a number of details which are explicitly indicated in the text.

We think that this book can be used as a textbook (it is actually based on lectures which I gave at the University of Jassy - Romania), and that it is also interesting for experts. We expect also that this book will be used by those high school teachers who are interested in the modernization of the teaching of geometry. More than this, we hope that it will satisfy a large group of readers (mathematically minded logicians, philosophers, engineers, etc.) to whom we tried to give a modern exposition about a subject which has been fascinating the human spirit for thousands of years and which is undoubtedly of great scientific interest.

Last but not least, let me acknowledge here my gratitude to all those people who helped me during the preparation of this book. I am mentioning first my former colleagues of the Chair of Geometry of the University of Jassy - Romania, for their helpful remarks about the Romanian version of this book (unfortunately, the publication of this version was interrupted at an advanced stage by Editura Technica, Bucarest, because I emigrated to Israel). Then, my thanks are going to Prof. R. Artzy (University of Haifa), for his revision of the English version, and to the University of Haifa and to the Merkaz Le Klita Be Mada - Israel for their financial support. The camera-ready copy was skilfully prepared by Mrs. Ana Burcat at the Sigma-Psi Scientific Typing Office in Haifa.

I. Vaisman

CONTENTS

FOUNDATIONS OF THREE-DIMENSIONAL EUCLIDEAN GEOMETRY

CHAPTER 0

INTRODUCTION

§1. THE AXIOMATIC METHOD AND ITS UTILIZATION IN EUCLIDEAN GEOMETRY

a. The Axiomatic Method

Generally, when speaking about the foundations of a mathematical theory, one understands that we have to discuss its starting point and the manner in which this theory is constructed. The requirement for a deep analysis of these problems comes from the fact that the mathematical theories are deductive theories, where logical reasoning is the main method of development.

Take, for example, elementary geometry, whose foundations are the object of the present book. Its development began in antiquity, first by a straightforward study of the properties of physical space, i.e., by an intuitive understanding of geometric figures. Then, the less elementary properties were obtained by proofs. After an accumulation of a large number of geometric theorems, it became clear that all these can be systematized starting with a small number of well chosen propositions. These initial propositions were called *axioms*. All the other results had to be obtained by logical deduction from the axioms. In this way, one arrived at the axiomatic construction of geometry. Of course, later on, the axiomatic method was improved and the axiom system of geometry was carefully worked out but, in principle, the method remained the same.

This line of development in elementary geometry is characteristic for all deductive theories and, particularly, for mathematical theories. We see thereby that, for such a theory, we must study the construction of the corresponding system of axioms and analyze its structure. All these constitute the foundations of the theory.

Let us give a few more details.

Elementary geometry was the first axiomatized theory. A corresponding system of axioms was given in the famous *Elements* of Euclid, which appeared in the third century B.C. In the *Elements* and, generally, during the period of the Greek antiquity and a long time afterward, the axioms were considered to be undoubted truths which we arrive at through our intuition. These axioms were expressed by using various notions and relations which we knew, again, by intuitive, descriptive definitions.

But such is no more the meaning of the axiomatic method today. After the mathematicians struggled for more than two thousand years with the famous problem of the parallels (see details in Chapter III), trying to prove an apparently unquestionable result (namely, that through any point not belonging to a given line there is just one parallel), and after finding the surprising solution that a geometry based on the contrary result exists, it became clear that looking at axioms as undoubted truths is unsatisfactory from the logical viewpoint.

Consequently, at the end of the 19th century, this conception was modified and a modern formulation of the axiomatic method constructed. Namely, one arrived at the idea of considering the *axioms* simply as a starting point or starting rules for a theory, whose truth did not have to be justified within the theory. These axioms employ some notions and relations which we introduce in a *purely formal* manner as initial symbols of the respective theory. Such symbols are denoted by words or signs and are called the *fundamental notions and relations* of the theory. The fundamental notions and relations do not have to be defined within the theory, which just means that they have a purely formal character. Any other notion or relation needed in the theory must be logically defined by the help of the fundamental ones and any result of the theory must be logically proven from the axioms. The results, once proven, are called *theorems*. All the previously considered elements, together, make up the *theory*.

The founder of the modern axiomatic method was David Hilbert (1862-1943) who among other things illustrated the method for the case of elementary geometry, by constructing for it a complete system of axioms, which today became the standard.

We saw thus which are the components of a theory, including the fundamental notions and relations, and the axioms. Together, these last three components form the so-called *axiom system* of the theory.

While the choice of the axiom system needs no justification within the theory (actually, because of this fact), it must be justified from the outside, in order to avoid *uninteresting theories*. This possible lack of interest may not only be due to the fact that the subject of the theory is unimportant (a danger which can be avoided by axiomatizing theories appearing in important mathematical and practical problems), but it is also possible to get a *logically uninteresting theory*. In fact, by an arbitrary choice of an axiom system it may happen that we obtain a *contradictory theory*, i.e., a theory where propositions exist which are true at the same time as their contraries. For obvious reasons, we are interested only in *non-contradictory theories*.

Hence, we have the last (but essential) argument requiring the analysis of a theory — called the *metatheory*.

Of course, the main problem of metatheory is the *non-contradiction problem* mentioned above. However, there are at least two more problems worth considering. First is the *independency problem*, asking us to inquire whether some proposition of a theory can or cannot be logically derived from a given system of other propositions. If it can be derived, it is called *dependent* and, if not, *independent*, on the respective system of propositions.

This is obviously an important problem, especially for the axioms, since it is desirable to use only independent axioms and to reduce as much as possible the number of the initially accepted axioms. It is also important to point out that the existence of an independent axiom yields on replacement by a contrary assumption a new theory, which might also prove to be interesting. For instance, the proof of the independence of the Euclidean axiom of the parallels (see Chapter III, §3, 4) led to the *non-Euclidean geometries*.

Finally, there is an important problem of knowing whether an axiom system, which gives an indirect definition of its fundamental notions and relations, is or is not uniquely determining its object.

The main instruments used in solving such problems are the so-called *models* of an axiom system. Namely, suppose we have a system A of axioms, defining a theory T, and suppose we also have another theory T' already analyzed. Then a *model* of A (or of T) within T' consists of a system of objects and relations of the theory T' and of a system of *interpretation rules* which allow to *represent* the fundamental notions and relations of A by the fixed objects and relations of T', in such a manner that the axioms of A become true propositions of T'.

Now, if T' is a non-contradictory theory, and if the system A of axioms has a model in T', then A (and hence the theory T generated by A) is also non-contradictory, because every contradiction generated by A would create a corresponding contradiction in T'. We thus have a method for studying the non-contradiction of a system of axioms, which consists of looking for corresponding models in other theories which are reasonably believed to be non-contradictory. If such a model exists we shall also say that A (and T) is *consistent*

— consistency being identified in this book with non-contradiction (though a more penetrating analysis can find differences between these two notions).

The models can also be used in studying the independence of the propositions within a theory. Actually, if we want to prove the independence of some proposition p with respect to a certain system of propositions $\{p_i\}$, we have to look for a model of the system $\{p_i\}$ where p does not hold. If such a model exists, p is independent of $\{p_i\}$ because, in the contrary case, p must hold true whenever the p_i are true. Let us also note that in the case when p is independent of the p_i, then, in our model, the contrary of p holds, such that the system consisting of $\{p_i\}$ and the contrary of p has to be consistent. We thus obtain a new consistent theory.

Finally, as for determining the object of an axiomatic theory, it is even the formulation of the problem which will be described by means of the models. Namely, two models of a system A of axioms are called *isomorphic*, if there is a one-to-one correspondence between them which preserves the structure described by the system A. The system A (and the theory T generated by A) is called *categorical* if all its models are pairwise isomorphic. It is clear that a categorical system of axioms defines uniquely, up to an isomorphism, its object, i.e., its model. For a noncategorical system there will exist non-isomorphic models. Let us note that both kinds of axiom systems are important: the categorical systems because they are uniquely defining their object, and the noncategorical systems because they give a general schema allowing the simultaneous study of several theories.

I would like to call the reader's attention to the fact that I do not intend to consider here the problem of analyzing the logic used in deductive theories. Though the problem is complicated but necessary for an in depth analysis of the subject, it is beyond the aim of this book.

b. Axiomatics of Euclidean Geometry

As we have already mentioned earlier, Hilbert gave a modern setting for the axiomatic method. Furthermore, he used this method in the development of elementary geometry and in solving the problems of constructing the corresponding theory and metatheory. Hilbert also gave a rigorous system of axioms, made an in depth study of the relations between the various axioms by conveniently grouping them together, and proved the consistency of the system of axioms by a model constructed within the arithmetics of real numbers. Hilbert also proved the independence of some important geometric propositions (for example, in [9] we find proof of the independence of the theorems of Desargues and Pappus-Pascal with respect to the plane incidence and order axioms, and the independence of the continuity axioms with respect to all the other geometrical axioms, which led to *non-Archimedian geometry*, etc.). Finally, Hilbert showed how from the geometric axioms we can deduce the *coordinate field* of the geometry by developing a theory of proportions and a calculus with geometrical segments [9].

Clearly, Hilbert based his work on earlier researches, but it is not my intention to dwell here on that history, which began with Euclid and culminated with such geometers of the 19th Century as Pasch, Peano, etc. But the main role and contribution can be attributed to David Hilbert, whose book *Grundlagen der Geometrie* (1855) became a classic and actually provided the solution of the problems of the foundations of elementary geometry.

In the book [9], mentioned above, elementary geometry is described as an axiomatic theory which starts with the fundamental notions of a *point*, a *line* and a *plane* and the fundamental relations of *belonging*, *betweeness*, *being congruent* and employs 20 axioms divided into five groups: 8 *incidence axioms*, 4 *order axioms*, 5 *congruence axioms*, 1 *axiom of the parallels* and 2 *continuity axioms*. Next, the main geometrical theorems are developed and the metamathematical problems are analyzed, as we described earlier in this text. The axioms are grouped according to fundamental relations which they describe, and the order of the groups of axioms is determined by logical requirements (some of the later axioms are formulated by means of the earlier ones) and by various other requirements, for example, Hilbert's interest in non-Archimedian geometry, etc.

Post-Hilbertian researches on the foundations of elementary geometry aimed either at giving a more complete construction of the theory or at introducing equivalent systems of axioms better suited for some specific aims.

As we have already pointed out, Hilbert constructs (by means of a calculus with segments) the field associated with geometry, which allows to define coordinates and to introduce analytical geometry. Actually, this can be done without using all the axioms of elementary geometry and, as a matter of fact, it can be done for the so-called *affine geometries*, where the axiom of the parallels holds but we have, generally, no congruence relation. It is also interesting that, conversely, the three-dimensional affine geometries are associated with fields (Chapter I, §4.) Today this problem is central which in turn implies a corresponding ordering of the system of geometrical axioms. Corresponding references are E. Artin [1] for the plane geometry, and C. Teleman [16, Chapter III] for three-dimensional geometry.

It is interesting to mention that many authors (beginning perhaps with H. Weyl [18]) decided to replace the geometric axioms with algebraic axioms and to develop geometry this way [14].

In this book, we consider the geometric development, i.e., essentially, we shall discuss Hilbert's axiomatics. But we shall always aim to determine all the geometries which satisfy the various groups of geometric axioms. In view of this aim, we shall introduce the axiom of the parallels as soon as possible, i.e., immediately after the incidence axioms. The group of incidence axioms remains the first group, and, naturally, the order axioms come third. For this last group, we consider a slight modification: we assume some *weaker* order axioms and a strengthened incidence axiom (namely, we asked for the existence of three and not two points on every line.) This is because the main properties of spaces based on incidence and order axioms only (which we call *geometrically ordered spaces*) hold with our weaker axioms as well, and by adding the axiom of the parallels, the whole system is equivalent to the usual one.

Furthermore, it is natural to place fourth the continuity axioms because, together with the other mentioned axioms, they define affine geometry over the field of real numbers [3,6].

Finally, the group of congruence axioms becomes the fifth group which characterizes the Euclidean geometries and, in particular, elementary geometry, which is Euclidean geometry over the real field. However, here again, we replace Hilbert's original formulation, with that of Kerékjártó [10], because it has the advantage of using only *congruence of segments*.

Many authors tried to base the metric character of Euclidean space on notions other than congruence of segments, for instance on the *scalar product* or on the perpendicularity of lines. However, in our opinion, the relation of congruence of segments is of a clearer geometrical significance, and we shall come back to this relation. This also allows for an elegant procedure of determining the fields associated with Euclidean geometries.

Generally, after introducing a new group of axioms, we study the geometric consequences which may be derived with or without the axiom of the parallels and, in the last case, we determine all the respective geometries by characterizing the corresponding field of coordinates. At the same time, we study the pertinent main problems of consistency, independence and categoricalness.

Due to the limitations which we have imposed upon the scope of the book, we cannot always give complete proofs. In some cases, generally simple, the proofs or their completion were deliberately left to the reader as exercises. This requires from the reader an active attitude toward the text and allows for a good understanding of the subject. In other situations, which are also explicitely indicated, complete proofs would be too long, while not introducing important new ideas (e.g., the orientability of the plane and the space based on incidence and order axioms only, polyhedra, congruence problems in space, etc.) and therefore were omitted without neglecting to inform the reader about the respective subjects.

Basically this book is self-contained. The cited bibliography and references are given to suggest further readings on the subject as well as to indicate the sources used in the book. The bibliography is selective and far from complete.

References used most are Hilbert [9], Kerékjártó [10] and Teleman [16, 17]. However, I am indebted to the many other listed references.

§2. USEFUL NOTIONS FROM OTHER MATHEMATICAL THEORIES

In order to make the book entirely self-contained, we shall insert here the most important notions of other mathematical disciplines which we shall use later. There is no need to read this section now; the reader may come back to it when needed.

a. Binary Relations

We admit that the reader knows the basic notions and results of elementary set theory. A binary relation on a set (or class) M is a law associating, by a well defined criterion, some pairs of elements of M. Defining the relation is the same thing as listing the set of the associated pairs, whence a binary relation on M may be identified with a subset of the cartesian product $M \times M$. Usually, one denotes by $a \rho b$ the fact that the elements $a, b \in M$ are associated by the relation ρ.

The binary relation ρ is *reflexive* if $a \rho a$, is *symmetrical* if $a \rho b$ implies $b \rho a$, and is *transitive*, if $a \rho b$ and $b \rho c$ imply $a \rho c$, all these for every $a, b, c \in M$. If all these three properties hold,

ρ is called an *equivalence relation*. Then, all the elements which are equivalent to $a \in M$ form the equivalence class of a, and this is denoted by $[a]_\rho$. M will be the union of the equivalence classes of its elements, and these are pairwise disjoint. The corresponding set of equivalence classes is denoted by M/ρ and is called the *quotient set* of M by the relation ρ. Defining new mathematical objects as equivalence classes of already known objects is an often used method and such definitions are called *definitions by abstraction*.

Two sets are called *equivalent* if there is a one-to-one* correspondence between them. This is an equivalence relation in the previous sense and its equivalence classes are called *cardinal numbers*. One can also say that two sets of the same class have the same *power*. The power of the set of natural numbers is called the *power of the countable* and that of the set of real numbers is the power of the *continuum*.

If a binary relation ρ is reflexive and transitive and if it is also *antisymmetric*, i.e., $a \rho b$ and $b \rho a$ imply $a = b$, the relation ρ is called a *partial ordering* and the pair (M,ρ) is a *partially ordered set*. If, moreover, any two elements of M are *comparable*, i.e., we necessarily have either $a \rho b$ or $b \rho a$, ρ is a *total* (*linear*) *ordering* and (M,ρ) is an *ordered set*. It may happen that such sets have a *first* (*smallest*) or *last* (*largest*) element.

We also note that there is an associated relation $\bar{\rho}$ defined by the fact that $a \bar{\rho} b$ iff $a \rho b$ and $a \neq b$. This is such that, $a \rho b$ iff $a \bar{\rho} b$ or $a = b$. Usually $\bar{\rho}$ is denoted by $<$ and ρ by $\leq$. The relation $\bar{\rho}$ is characterized by transitivity and by the property of *trichotomy*, which means that, for every $a,b \in M$, one and only one of the situations $a \bar{\rho} b$, $a = b$, $b \bar{\rho} a$ holds. An ordered set is called *dense* if we have always at least one element between two given elements of the set. If M_1 and M_2 are two ordered sets and $f : M_1 \to M_2$ is a bijective correspondence which preserves the order (i.e., $a \leq b$ implies $f(a) \leq f(b)$), f is called a *similarity* and M_1, M_2 are called *similar sets*.

* In this book "one-to-one" and "bijective" are the same.

Finally, let us mention that for every order relation ρ there is an *inverse order* ρ^* defined by $a\,\rho^*\,b$ if $b\,\rho\,a$. If we denote ρ by $\leq$, ρ^* is usually denoted by $\geq$.

b. Groups

A law which associates with some elements of M (or of more sets) a well defined other element is called an *algebraic operation*. A set (or more) endowed with a number of algebraic operations which satisfy various axioms is called an *algebraic structure*. A map between two sets endowed with similar algebraic structures, which preserves the algebraic operations, is called a *homomorphism* and, if it is also one-to-one, it is called an *isomorphism*. When the two structures coincide, the homomorphisms are called *endomorphisms* and the isomorphisms are called *automorphisms*. An injective homomorphism is called a *monomorphism* and a surjective one is called an *epimorphism*.

A set M, endowed with a binary algebraic operation (usually called a *product*) associating with every pair $a,b \in M$ an element $c \in M$, denoted by $c = ab$, is called a *semigroup* if the operation is *associative*, i.e., if we have

$$a(bc) = (ab)c$$

for every triple of elements of M. If we also have the property of *commutativity*, i.e. $ab = ba$ for every $a,b \in M$, the semigroup is called *commutative* or *abelian*. If there is an $e \in M$ such that $ae = ea = a$ $(a \in M)$, this is called a *unit element* and the semigroup is *with unit*. One often denotes e by 1. In a semigroup with unit, an element a such that there is an element a^{-1} with $aa^{-1} = a^{-1}a = e$ is called *invertible* and a^{-1} is the *inverse* of a. A semigroup with a unit and having only invertible elements is called a *group*. In particular, when we have *abelian groups* we usually call the operation a *sum* and denote it by +, while e is denoted by 0 and the inverse, now called the *opposite*, is denoted by $-a$.

For instance, if A is an arbitrary set and if we denote by $L(A)$ the set of all formal finite linear combinations of the form $\Sigma n_i a_i$, where n_i are integers and $a_i \in A$ (these combinations are formal in the sense that they are defined only as new symbols), then, defining the sum as for polynomials, we see that $L(A)$ becomes an abelian group, which is called the *free abelian group* generated by the set A.

If G is a group and G' a subset of G, G' is called a *subgroup* if it is also a group with respect to the operation defined like in G. This happens if G' is closed with respect to the operation of G and to the construction of the inverse element. If, moreover, G' is such that, for every $g \in G$ and $g' \in G'$, $gg'g^{-1} \in G$, it is called a *normal* or *invariant subgroup*. In this case, the relation $g_1 \sim g_2$ if there is $g' \in G'$ with $g_1 = g_2 g'$ (or, equivalently, $g_2^{-1} g_1 \in G'$) is an equivalence relation and the quotient set, denoted here by G/G', has an induced group structure. Namely, an element of G/G' is a class $g\,G'$ (consisting of all the products gg' with $g' \in G'$) and, if we define the product of two classes by $(gG')(\bar{g}G') = (g\bar{g})G'$, we can see that it is well defined and gives G/G' the announced group structure. G/G' is called the *quotient group* of G by G'. If G is abelian, all its subgroups are normal and are defining corresponding quotient groups.

Let G be an arbitrary group and $a,b \in G$. Then $aba^{-1}b^{-1}$ is called the *commutator* of the pair (a,b). It is easy to see that the subset $K \subseteq G$ consisting of all the finite products of commutators is an invariant subgroup of G, namely, it is the smallest invariant subgroup which defines an abelian quotient group. This K is called the *commutator subgroup* of G.

Let $f : G_1 \to G_2$ be a homomorphism of two groups. Then we can introduce the sets

$$\ker f = \{g \in G/f(g) = e_2\}, \quad \operatorname{im} f = f(G_1) \subseteq G_2 \quad ,$$

where e_2 is the unit of G_2. One sees that $\ker f$ is an invariant subgroup of G_1, and we call it the *kernel* of f, and $\operatorname{im} f$ is some subgroup of G_2 called the *image* of f. It is easy to prove that f is a monomorphism iff $\ker f = e_1$ (the unit of G_1) and it is an epimorphism iff $\operatorname{im} f = G_2$. Both conditions hold iff f is an isomorphism. A basic algebraic theorem says that for every homomorphism f, $G_1/\ker f$ is isomorphic to $\operatorname{im} f$.

Most often, the groups encountered in geometry are transformation groups. If G is a group and M a set, a *left action of G as a transformation group on M* is a mapping $a : G \times M \to M$ with the following two properties:

$$1.\ a(e,m) = m\ ; \qquad 2.\ a(g_1 g_2, m) = a[g_1, a(g_2, m)]\ ,$$

where $m \in M$, $g_1, g_2 \in G$ and e is the unit of G. (We can define analogously right actions.) Instead of $a(g,m)$ we shall denote simply gm and, for a fixed g, this defines a transformation $M \to M$, which explains the above terminology. For instance, we have the group of all bijective transformations of a set M onto itself, where the group operation is a usual composition of mappings. We can also consider various subgroups of it. An action of G on M is called *effective* if $gm = m$ for every $m \in M$ implies $g = e$. In the opposite case, the action is not effective and then the set of the elements $g \in G$, with $gm = m$ for every m, is a normal subgroup G' of G, such that G/G' has an induced effective action on M. If in the previous discussion we replace "for every m" with "for at least one m," the term "effective" has to be replaced with "free" and then we have or do not have a *free action*. Generally, if $gm = m$, m is called a *fixed point* of the transformation g.

c. Fields and Linear Spaces

Another fundamental algebraic structure is the *ring structure*. A *ring* is an additive abelian group A, where a second operation called *product* is also defined such that

$$a(b+c) = ab+ac, \quad (b+c)a = ba+ca \quad ,$$

i.e., the product is *distributive* with respect to the sum.

If, moreover, the product is associative, commutative or has a unit element, we have, correspondingly, an associative ring, a commutative ring, a unitary ring, etc.

If the set of non-zero (0 is the unit of the sum) elements of the ring A is a group with respect to the product, A is called a *field* and if the product is also commutative, we have a *commutative field*. Any field has at least two elements: 0 and 1 which are respectively the units of the sum and of the product.

In rings and fields the usual rules for the signs + and - hold. As a matter of fact, the field is just the abstract algebraic structure where the classical four arithmetic operations (addition, subtraction, multiplication and division) may be performed and have the usual properties. We can see that there are both commutative fields, e.g., the rational, real or complex numbers, and non-commutative fields, e.g., the field of the *quaternions*, which are formal expressions of the type

$$\alpha = a + bi + cj + dk ,$$

where i, j, k are symbols which have to be multiplied by the following rules: $i^2 = j^2 = k^2 = -1$, $ij = -ji = k$, $jk = -kj = i$, $ki = -ik = j$. The sum and product of quaternions are then defined like for polynomials.

It is important to define the *characteristic* of a field K. This is the smallest natural number $n \neq 0$ such that $na = 0$ for $a \in K$, if such numbers exist, and 0 if such n do not exist. E.g., the classes of integers modulo p form a ring, which is a field if p is prime and it has the characteristic p. Particularly, the integers modulo 2 form a field of just two elements 0 and 1, and this field is obviously of characteristic 2.

Let A be a ring and I a subset of A. Then I is a *subring* if it is closed with respect to sums and products in A, and it is also a ring for the induced operations. If, moreover, for every $a \in A$ and $i \in I$, $ai \in I$ (which is called the left absorption property), I is called a *left ideal* in A. Analogously we define *right absorption* and *right ideals*. If I is both a left and a right ideal, we call it simply an *ideal* of A. In this last case, the abelian quotient group becomes a ring if we define the product of two classes to be the class of the product of two elements in these classes. We shall denote this new ring by A/I and call it the *quotient ring*.

If $f : A_1 \to A_2$ is a ring homomorphism, ker f and im f are naturally defined by the corresponding abelian group structures. It is easy to see that ker f is an ideal of A_1 and im f is isomorphic (as a ring) with $A_1 \mid \ker f$.

In particular, a field K has but two ideals, K itself and 0, because, if I is an ideal of K and $a \neq 0$ is an element of I, $a^{-1}a = 1 \in I$ (in view of the absorption property), and then $b = b.1 \in I$ for every $b \in K$, which means $I = K$. It follows that for a homomorphism $f : K_1 \to K_2$ of fields we must have either ker $f = 0$ and f is injective, i.e., a monomorphism, or ker $f = K_1$ and f sends every element of K_1 to $0 \in K_2$. Hence, if $f : K \to K$ is an epimorphism (surjective homomorphism) f is necessarily an automorphism of the field K.

Next, let K be an arbitrary field, not necessarily commutative and V be an additive abelian group (of course, there is no relation between the sum operations of K and V, which, however, are both denoted by +). V is said to be a *left linear* or *vector space* over the field K, if there is provided an operation $K \times V \to V$, associating with every $k \in K$ and $\bar{v}$ in V an element $k\bar{v} \in V$ such that

1. $1\bar{v} = \bar{v}$;
2. $k_1(k_2\bar{v}) = (k_1k_2)\bar{v}$;
3. $(k_1 + k_2)\bar{v} = k_1\bar{v} + k_2\bar{v}$;
4. $k(\bar{v}_1 + \bar{v}_2) = k\bar{v}_1 + k\bar{v}_2$.

In this case, the elements of K are called *scalars*, those of V *vectors* and the above operation is the *multiplication of a vector by a scalar*. We can define analogously a *right linear space* while "left" and "right" coincide if the field K is commutative.

In the theory of linear spaces, the fundamental notion is that of *linear independence* of vectors. Considering the previous notation, the vectors $\bar{v}_1,\dots,\bar{v}_n$ are *linearly independent* if any relation $\sum_{i=1}^{n} k_i \bar{v}_i = \bar{0}$ ($\bar{0}$ is the unit of the additive structure of V) implies $k_i = 0$ for all $i = 1,\dots,n$. If there is at least one relation of this type, with at least one $k_i \neq 0$, the vectors $\bar{v}_i$ are called *linearly dependent*.

If there is some natural number n such that: (1) there is at least one system of n independent vectors; (2) any system of $m > n$ vectors consists of linearly dependent vectors, then n is uniquely defined and it is called the *dimension* of V. If such an n does not exist, V is *infinite-dimensional*. If V has the dimension n, any system of n linearly independent vectors is called a *basis* of V. Let $\bar{e}_i$ $(i = 1,\dots,n)$ be such a basis. Then, for every $v \in V$, $(\bar{v},\bar{e}_i)$ are linearly dependent vectors, and we get some relation

$$\bar{v} = \sum_{i=1}^{n} x^i \bar{e}_i \quad ,$$

where the x^i are uniquely defined and they are called the *coordinates* of $\bar{v}$ with respect to the basis $\{\bar{e}_i\}$.

It is now obvious that for two bases $\{\bar{e}_i\}$, $\{\bar{e}_i'\}$ we have some connection formulas of the form

$$\bar{e}_i' = \sum_{j=1}^{n} p_i^j \bar{e}_j \quad , \quad \bar{e}_i = \sum_{j=1}^{n} q_i^j \bar{e}_j' \ ,$$

where (p_i^j) and (q_i^j) are matrices inverse to each other. This allows for finding the relations between the coordinates of a vector in two different bases (Ch. I, §4).

In a similar manner as for groups or rings, we may define the notion of a *linear subspace*. If V is a left n-dimensional linear space over K, and $V' \subseteq V$ is a linear subspace, V' is clearly finite dimensional as well. Hence V' must be a subset consisting of the vectors of the form $\sum_{i=1}^{h} k_i \bar{v}_i$ where $\bar{v}_1, \ldots, \bar{v}_h$ are linearly independent vectors defining a basis for V'. Of course, $h \leq n$ is then the dimension of V'.

Any additional and specific algebraic information to be used in this book will be provided only when needed. For a complete treatment of the subjects mentioned in this section, see, for example [11].

d. Topological Spaces

Another class of fundamental mathematical structures are the *topological structures*. Consider a set T and suppose that a family τ of subsets of T is fixed such that

(1) ϕ(the empty set) and T belong to τ;

(2) if $U_a \in \tau$ (where α belongs to an arbitrary set of indices) then $\bigcup_a U_a \in \tau$;

(3) if $U_i \in \tau$ $(i = 1,2)$, then $U_1 \cap U_2 \in \tau$.

Then, τ is called a *topology* of T, the sets of τ are called *open sets*, the corresponding complementary sets are called *closed sets* and the pair (T, τ) is called a *topological space*. When no confusion is possible, the topological space is denoted simply by T.

Let T be a topological space and x a *point* (element) of T. Then, a *neighborhood* of x is any subset of T which contains an open subset containing x. Often, one uses only the *open neighborhoods*.

If $M \subseteq T$, the *interior* $\overset{\circ}{M}$ of M is the largest open subset of T contained in M. The points of $\overset{\circ}{M}$ are called *interior points* and they are characterized by having an entire open neighborhood contained in M. Using the same notation, the smallest closed set containing M is called the *closure* of M and is denoted by $\overline{M}$. Its points are called *adherence points* and they are characterized by the property that every neighborhood of such a point has a non-void intersection with M. If, moreover, every neighborhood of the point has common points with M, which are different from the given point, the last one is called a *limit* or *accumulation point* of M. The set of the limit points of M is the *derived set* M' of M. Finally, $\overset{\circ}{M} = \overline{M} - \overset{\circ}{M}$ is called the *boundary* of M and its points are *boundary points*.

If $M \subseteq T$ and $\overline{M} = T$, M is said to be *everywhere dense* in T. If a space T has a countable subset M which is everywhere dense in T, T is called a *separable space*.

Generally, to define a topology on a set T, it suffices to have only part of the system τ, namely, it is enough to give $\tau' \subset \tau$ such that every open set will be a union of sets in τ'. Such a subsystem τ' of τ is called a *basis* of the topology τ. Moreover, in view of the definition of a topology, it suffices to define a subsystem τ'' such that every open set be an arbitrary union of some finite intersections of sets in τ''. Such a τ'' is called a *subbasis* of τ. As a consequence, it follows that every family of subsets of a given set T is the subbasis of a well defined topology of T, whose open sets will be ϕ, T and the unions of the finite intersections of sets of the family.

Similarly, in order to perceive all the neighborhoods of a point $x \in T$ it suffices to know a family $\{V_x\}$ of open neighborhoods of x such that any neighborhood of x contains an element of this family. Such a family is called a *basis of open neighborhoods* of x. Clearly, if we have such a basis for every $x \in T$, the topology is defined by the condition that a set D is open if $x \in D$ implies $U_x \subseteq D$ for some open neighborhood U_x of x (Of course, the bases of neighborhoods of the points are not arbitrary families of sets. They must satisfy some conditions which follow from the definition of a topology.)

Let T be a topological space and let M be a subset of T. Then, it is easy to see that, by defining the open sets in M as the intersections of M with the open sets of T, we get a topology on M. This is called the *relative* or *induced topology* and M endowed with this topology is called a *topological subspace* of T. The neighborhoods of the induced topology are also the intersections of M with the neighborhoods of the topology of T.

A mapping $\varphi : T_1 \to T_2$ between two topological spaces is called *continuous* if the preimage of any open subset of T_2 by φ is an open subset of T_1. If, moreover, φ is bijective and φ^{-1} is also continuous, φ is called a *homeomorphism* and the two spaces are called *homeomorphic*.

There are many important special classes of topological spaces and we shall consider here a few of them. The topological space T is *separate* or *Hausdorff* (do not confuse it with *separable*), if for any two different points of T, there are disjoint neighborhoods of these points. The space T is *compact* if it is Hausdorff and whenever $T = \cup_a U_a$, where $\{U_a\}$ is some family of open sets of T (called then a *covering* of T), there is a finite number of the sets U_a, whose union is T (i.e., any open covering of T contains a finite covering.) Finally, the space T is called *connected* if there is no decomposition $T = D_1 \cup D_2$, where D_1 and D_2 are non-void open subsets of T.

For a complete development regarding the topological spaces see, for instance [4].

It would also be useful for the reader to recall the elementary analytic geometry and the definition of the number systems (e.g., [7]), especially that of the real numbers R including the topology defined on R by the open intervals and in the space $R^n = \{(x^i)/i = 1,\ldots,n;\ x^i \in R\}$ by the open cubes (which may be found in any book on Calculus, e.g., [8].)

CHAPTER 1
AFFINE SPACES

§1. INCIDENCE AXIOMS AND THEIR CONSEQUENCES

The final aim of this book is to construct a mathematical model of physical space, called the *three-dimensional Euclidean space*, and whose study is called *elementary geometry*. We assume that we already have an intuitive description of this geometry, which is used as a heuristic support of the present axiomatic construction.

Consequently, we choose as a fundamental notion the notion of a *point* where the points form a set called *space*. The space has a structure defined by two systems of subsets called *lines* and *planes*, respectively. The notions of a line and a plane are also considered to be fundamental (in the sense of Section 1 of the Introduction, of course). We shall denote the points by capital Latin letters, the lines by small Latin letters and the planes by Greek letters. The space will be denoted by S. (It is worthwhile noting that often, one does not consider from the beginning that the lines and the planes are sets of points [9,10]).

A first chapter of elementary geometry must clearly deal with the various inclusion properties (in the sense of set theory) of points, lines, and planes, which in geometrical language are called *incidence properties*. First, we shall formulate a number of *incidence axioms*, which are characteristic of the considered fundamental objects. These form the first group of axioms of elementary geometry and they are:

I 1. Two distinct points belong to one and only one line.

I 2. Any line contains at least two distinct points.

I 3. There are three points which do not belong to the same line.

I 4. Every plane contains at least one point, and three points, not on the same line, belong to one and only one plane.

I 5. If two points of a line belong to some plane, every point of that line belongs to this plane (the line is included in the plane).

I 6. If two planes have a common point, they have at least one more common point.

I 7. There are four points which do not belong to the same plane.

When speaking about incidence, we shall use the classical language of intuitive elementary geometry. Let us remark that the axioms I 1-I 3 are, actually, plane axioms, and, in order to study plane geometry (where we have a set of points and lines defining the plane), we have to consider only these three incidence axioms.

It is easy to prove the consistency of the incidence axioms. In fact, if the points are any four objects, the lines are the pairs of these objects, and the planes are any triples of objects, propositions I 1 through I 7 clearly hold. Thus, we have a finite and obviously noncontradictory model of the considered axioms which are thereby consistent.

For all the questions raised in this and the following section the formulated axioms suffice, but some strengthening of them will be required later. Namely, we shall need to replace I 2 by

I 2'. Any line contains at least three distinct points.

The incidence axioms thus obtained are called the *strong incidence axioms* and their use will always be explicitly indicated.

From the incidence axioms, we can get consequences regarding the determination of points, lines and planes and the possible positions of these. Based on some examples presented in the text, the reader can proceed to prove by himself any other similar proposition he may need.

1. THEOREM. For any line, there is at least one point not on this line. For any plane, there is at least one point not on this plane.

In fact, let d be a line and α a plane. Then S-d contains at least one of the points of I 3 and $S - \alpha$ contains at least one of the points of I 7.

2. THEOREM. Given a point A and a line a which does not contain this point, there is a unique plane containing A and a. Two concurrent lines define a unique plane which contains them.

In fact, given $A \bar{\in} a$, take two points $B, C \in a$ (I 2) and consider next the plane ABC given by I 4. The uniqueness follows from the fact that any plane containing A and a contains A, B and C whence, by I 4, it must be just the plane obtained before. The second assertion of the theorem follows in a similar manner.

3. THEOREM. If two planes have a common point, they actually have a common line.

This follows trivially from the Axioms I 5 and I 6.

4. THEOREM. Any plane contains at least three noncollinear points.

Consider a plane α. By Axiom I 4 it has a point A and, by Theorem 1, there is a point $M \bar{\in} \alpha$. Using again Theorem 1, we may define

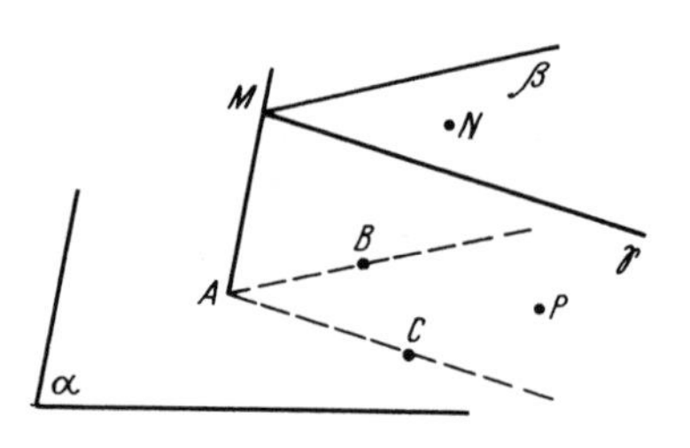

Fig. 1

a point N which does not belong to the line AM (defined by Axiom I1). Then, by Theorem 2, we have a plane β containing A,M,N and by Axiom I6, we have one more point B in $\alpha \cap \beta$. Similarly, using a point P not in β and the plane γ defined by A,M,P we finally get the third desired point C. A, B, and C must be noncollinear to avoid a contradiction (See Fig. 1).

Theorem 4 expresses the *two-dimensional character* of the planes. It is obviously equivalent to the fact that *any plane has at least two concurrent lines.*

5. THEOREM. Two lines are always in one and only one of the following situations: a) coincident, b) concurrent, c) without common points and not in the same plane, d) without common points and in the same plane.

In fact, taking as the classifying criterion the number n of the common points of the two lines, we meet the cases $n \geq 2$, $n = 1$, $n = 0$, which give just the mentioned situations.

Related to this theorem, we have

6. DEFINITION. Two lines in position c) of Theorem 5 are called *skew lines*, and two lines in position d) of Theorem 5 are called *parallel lines*; the symbol of parallelism is $\parallel$.

It is important to note that Theorem 5 is a theorem of classification but not of existence, i.e., it does not assure the *existence* of pairs of lines of the mentioned types. Particularly, it does not answer the important question of the existence of parallel lines. Moreover, we may see that from the incidence axioms only we cannot derive this answer. In fact, in the previously considered model of the incidence axioms, consisting of four objects, parallel lines, obviously, do not exist, while in the intuitive model of elementary geometry, for instance, such lines do exist. In other words, the existence of parallel lines is independent (in the sense of the Section 1 of the Introduction) of the incidence axioms.

In the remainder of this section, we assume the convention that any assertion about parallel lines is to be considered only when such lines exist, and that it is meaningless otherwise.

7. PROPOSITION. Consider $A \bar{\in} a$. Then the plane which they define (by Theorem 2) is the set of all points located on the lines which contain A and are either concurrent or parallel to a. Consider next two concurrent lines a,b. Then the plane which they define (by Theorem 2) is the set of all points located on the lines which meet a and b at distinct points or meet either a or b and are parallel with b or a respectively.

The proof (by double inclusion) is left as an exercise for the reader.

8. THEOREM. A line a and a plane α are in one and only one of the following situations: a) $a \subset \alpha$; b) concurrent (one point in common); c) they have no common point.

The proof is like that of Theorem 5.

9. DEFINITION. A line and a plane without common points are called *parallel*.

10. THEOREM. Let d be a line which is not contained in the plane α and is parallel with some line d' of α. Then $d \parallel \alpha$.

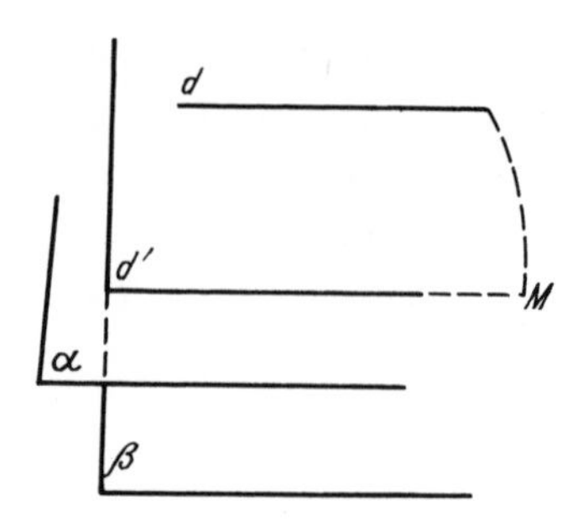

Fig. 2

In fact, the parallel lines d and d' define a plane β (Fig. 2) which, in view of I4 and I6, has in common with α the line d'. Hence, if d would intersect α in M, then $M \in \alpha \cap \beta = d'$, in contradiction with the hypothesis $d \parallel d'$.

11. THEOREM. Two planes are in one and only one of the following situations: a) coincident; b) concurrent (having a line in common); c) without common points.

Once again, the proof is that of Theorem 5, but it is based on Axioms I4, I6, and on Theorem 3.

12. DEFINITION. Two planes without common points are called parallel planes.

§2. THE AXIOM OF THE PARALLELS AND ITS INFLUENCE ON THE INCIDENCE PROPERTIES

a. Affine Spaces

We have already noted that the theory of the parallel lines requires a new axiom. This will form the second group of our system of axioms and will be called the *axiom of the parallels* or, shortly, the *parallel axiom*. By various choices of the parallel axiom one gets various geometries and, since our aim is the study of Euclidean geometry, we shall accept the *Euclidean parallel axiom* in the following form

II. If A is any point and a is any line such that $A \bar{\in} a$, there is one and only one line b containing A and parallel to a.

Let us note, in passing, that other parallel axioms used are: the axiom of Lobachevski asserting that *through a point not on a line there are at least two parallels to that line*, and Riemann's axiom asserting that *there are no parallel lines*. These axioms lead, respectively, to the geometries of Lobachevski and Riemann, which are as consistent as the Euclidean geometry because they do have Euclidean models. In fact, we get a Euclidean model of the Lobachevski plane by taking, for instance, as points the interior points of a Euclidean circle and as lines the chords of this circle (see also Ch. III §4). A model of the Riemann plane is obtained by taking as points the pairs of diametrically opposite points of a Euclidean sphere, and the lines being defined by means of the great circles of this sphere.

1. DEFINITION. The spaces for which the axioms I1-I7 and II hold are called *affine spaces*.

In affine spaces new incidence properties are to be expected and we shall consider some of them in this section.

2. THEOREM. Let A be a point of the plane α and d a line parallel to α. Then, the parallel to d through A is contained in α.

Let $d' \parallel d$ and $A \in d'$ (Fig. 3). Consider the plane β passing through A and d and denote $d'' = \alpha \cap \beta$. $d \parallel \alpha$ implies $d \parallel d''$ and, because of Axiom II, $d'' = d'$, q.e.d.

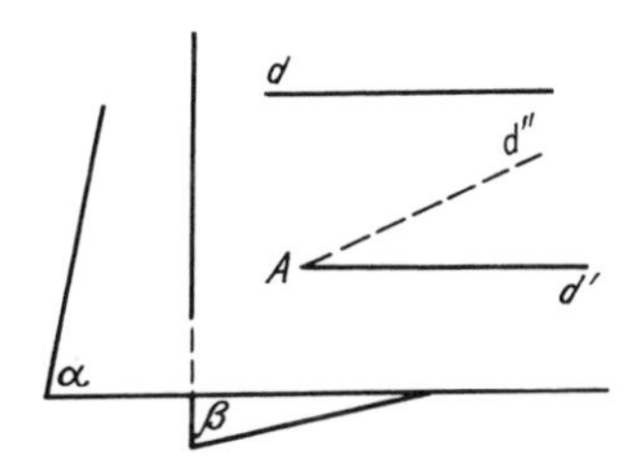

Fig. 3

By combining this result with Theorem 1.10 we obtain

3. THEOREM. A line not contained in a plane is parallel to that plane iff it is parallel with a line of that plane.

The following, more general, result will also be useful.

4. PROPOSITION. If $A \in \alpha$ and d is a line through A which is parallel to the plane $\beta \parallel \alpha$, then d is in the plane α.

In fact (Fig. 4), by Theorem 3, we have some line $d' \parallel d$ which is contained in β. To avoid a contradiction, we must have $d' \parallel \alpha$ and the proposition follows by applying Theorem 2 for d.

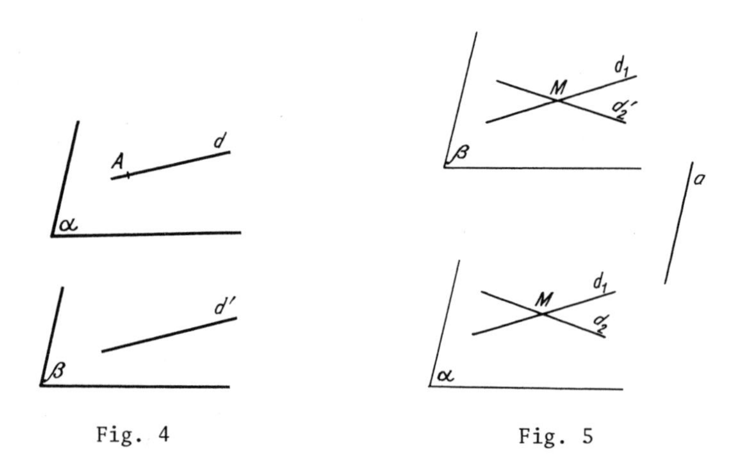

Fig. 4 Fig. 5

5. THEOREM. Through any point not situated in a plane there is a plane and only one parallel to the given plane.

Denote by M the point and by α the plane (Fig. 5), and take in α two concurrent lines d_1 and d_2 (which exist by Theorem 1.4). Next, consider through M, the lines $d_1' \parallel d_1$, $d_2' \parallel d_2$, which define a unique plane β. Now, suppose that there is a line $a = \alpha \cap \beta$. Then a meets neither d_1 nor d_2 because these lines are parallel to β (by Theorem 3). Hence, $a \parallel d_1$ and $a \parallel d_2$, which contradicts the Parallel Axiom. It follows that α and β have no common points, which proves the existence of the parallel plane required. Next, if γ is any plane through M which is parallel to α, it must contain d_1' and d_2', by Proposition 4. Hence, $\gamma = \beta$, and the uniqueness is also proven.

Let us also note that, in view of Proposition 4, the previous plane β is the locus of all the lines containing M and parallel to α.

A simple but important result is

6. THEOREM. Two planes which are parallel to a third plane are parallel planes.

This is a straightforward consequence of Theorem 5.

It is less simple, but possible, to prove a similar result for lines.

7. THEOREM. Two lines which are parallel to the same third line are parallel lines.

In fact, consider the lines a,b,c such that $b \parallel a$, $c \parallel a$ (Fig. 6). If they belong to the same plane, the result is a trivial consequence of the parallel axiom. If this does not happen, let's take a point $A \in c$ and consider the distinct planes α defined by A,b and β defined by a,c. These planes have a common line d which contains A. According to Theorem 3, $b \parallel \beta$ and $a \parallel \alpha$, whence it follows, by reductio ad absurdum, that $d \parallel b$ and $d \parallel a$. Now, by the parallel axiom, $d \parallel a$, $c \parallel a$ implies $d = c$, and the relation $d \parallel b$ becomes $c \parallel b$, q.e.d.

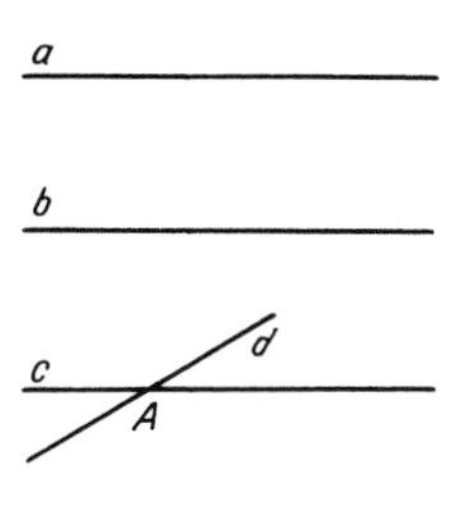

Fig. 6

The proof of the next theorem is left as an exercise to the reader.

8. THEOREM. If for the planes α, β, γ we have $\alpha \parallel \beta$ and γ concurrent with α, then γ is concurrent with β and the lines $\gamma \cap \alpha$, $\gamma \cap \beta$ are parallel.

b. Projective Spaces

Theorem 7 leads to a very important geometrical construction, namely, that of the *enlarged affine* or *projective space* and, though we do not intend to study projective geometry, it is worthwhile to describe this construction.

Denote by $\underline{\parallel}$ the binary relation which associates two lines which are either coincident or parallel. Theorem 7 together with the definition of parallelism show that this relation is an equivalence relation, which allows us to state the following

9. DEFINITION. An *improper point* of the affine space is a class of lines equivalent by the relation $\underline{\parallel}$. The union of the affine space and of the set of its improper points is a new set called the *enlarged affine space*.

In the enlarged affine space we have, as important subsets, the *enlarged lines* obtained by adding to every usual line the improper point defined by the equivalence class of that line. Also, we have the *enlarged planes* defined by adding to each usual plane all the improper points defined by the lines of the plane. The set of all improper points of an enlarged plane is called an *improper line*. It is obvious that two parallel planes define the same improper line, whence we see that the improper lines might also be defined as the equivalence classes of the relation $\underline{\parallel}$ for planes. Finally, the set of all improper points of enlarged affine space is called the *improper plane* of the space. In this context, the usual points, the usual enlarged lines, and the usual enlarged planes of affine space are labelled as *proper*.

We have thus obtained for the enlarged affine space a structure of enlarged lines and planes, which may be proper or improper, and if we study common properties of all these objects, either proper or improper, we get the so-called *projective geometry*, and the space is then called *projective space*.

The incidence properties of projective space are stated by

10. THEOREM. For the enlarged affine space, all properties of the strong incidence axioms hold. Moreover, any two distinct lines which lie in the same plane have a common point and any two distinct planes have a common line.

As a matter of fact, these properties have now a richer content because they are true for the enlarged and the improper elements. The second part of the theorem shows that parallelism does not exist in projective geometry.

The proof of the properties in Theorem 10 is carried out by considering all possible cases. For instance, to obtain I1, i.e., the property that two distinct points belong to a unique common line, we proceed as follows. First, if $A \neq B$ are proper points they belong by I1, to a unique proper line and belong not to any improper line (because those have only improper points).

Next, suppose A is proper and B improper. Then, the enlarged line, defined by the parallel through A to the class defining B, is clearly the unique line containing A and B.

Finally, if both A and B are improper, they belong to the improper line defined by the planes which are parallel to the lines defining A and B. Also, this is the unique line containing A and B (in particular, A and B do not belong to a proper line because such a line has only one improper point).

Further, I2' follows from I2 and from the existence of the improper point, for a proper line. For an improper line, consider a plane, defining the line and take two usual concurrent lines in this plane. The improper points of these lines define two points on the given improper line. A third point is, for instance, the improper point of the diagonal of any parallelogram with two sides on the chosen concurrent lines, etc.

Similarly, in order to prove that two distinct coplanar lines, a,b have a common point, we consider first the case when a,b are proper and concurrent, and the result is trivial. Next, we consider the case when a,b are enlarged lines defined by proper parallel lines, where a and b have a common improper point. Then, let a be proper and b improper; in this case, they will have as a common point the improper point of a. Finally, if both a and b are improper, their common point is the improper point of the intersecting line of two planes defining a and b.

In the same manner, one proves all the properties in Theorem 10 and we leave their proof as an exercise to the reader. We also leave to the reader the proof of any other similar incidence property we may eventually need.

c. Desargues Theorems

Now, we shall study some very important incidence properties known as the *direct and converse Desargues Theorems*, whose general form belongs to projective geometry, i.e., to the enlarged affine space and holds for both proper and improper elements. For the sake of convenience let us give first:

11. DEFINITION. We are calling *segment* a pair of distinct points called the *ends* of the segment. A *triangle* is a triple of noncollinear points called the *vertices* of the triangle. The segments defined by the vertices are called *sides* and the lines defined by the vertices are also called *sides*. (The double use of the word *side* causes no confusion in the sequel.)

Now, we may proceed to formulate for projective space.

12. THEOREM (Desargues). Let ABC and $A'B'C'$ be two triangles. If the lines AA', BB', CC' are concurrent at O, the points $P = AB \cap A'B'$, $Q = BC \cap B'C'$, $R = AC \cap A'C'$ are collinear and conversely.

In other words, if there is some one-to-one correspondence between the vertices of two triangles, such that the lines joining corresponding vertices are concurrent, then the corresponding sides of these triangles have collinear intersection points and conversely.

We shall first prove the theorem in the case where the two triangles belong to distinct planes α and α'. Then (Fig. 7), AB and $A'B'$ belong to the same plane OAB and, by Theorem 10, have the common point P which is in α and α' whence it is on the intersection line $\alpha \cap \alpha'$ (whose existence follows again by Theorem 10). Similarly, we establish the existence of the points Q and R which also belong to $\alpha \cap \alpha'$, whence the assertion of the theorem follows.

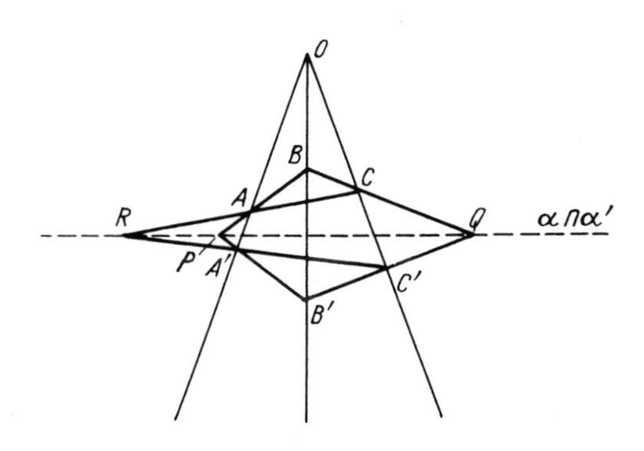

Fig. 7

Let us go now to the case of two triangles which lie in the same plane α (Fig. 8). Then, let O' be a point which is not in α and O'' a third point on the line OO'. By considering the planes containing OO', we can easily establish the existence of the points

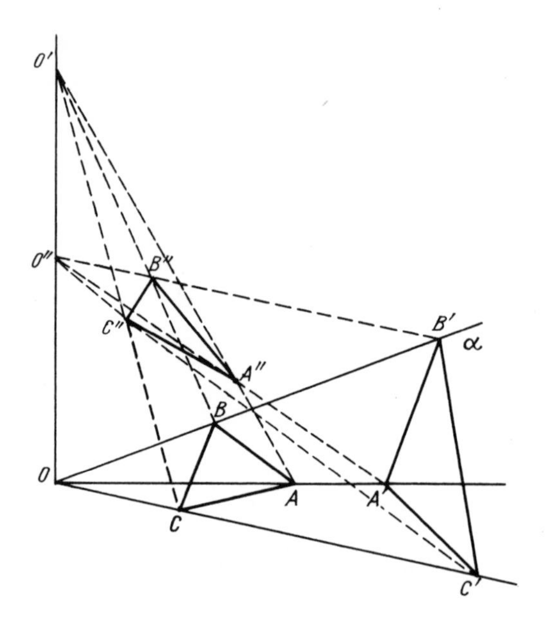

Fig. 8

$A'' = O'A \cap O''A'$, $B'' = O'B \cap O''B'$, $C'' = O'C \cap O''C'$, which lie in some plane $\beta \neq \alpha$. $\alpha \cap \beta$ is then a well defined line d.

By applying the proven part of the theorem to the pairs of triangles $A''B''C''$, ABC and $A''B''C''$, $A'B'C'$, it follows that on the line d we have the intersection points

$$R' = A''C'' \cap AC, \quad Q' = B''C'' \cap BC, \quad P' = A''B'' \cap AB \ ,$$

$$R'' = A''C'' \cap A'C', \quad Q'' = B''C'' \cap B'C', \quad P'' = A''B'' \cap A'B' \ .$$

Because, for instance, $A''C''$ is not in α and R',R'' are, both, in $A''C''$ and in α, we get $R' = R''$ and, similarly, $Q' = Q''$ and $P' = P''$. It is then obvious that $R' = R'' = R$, $Q' = Q'' = Q$, $P' = P'' = P$, where P,Q,R are the points of Theorem 12, and that $P,Q,R \in d$, which completes the proof of the direct part of the theorem.

The converse result has the same proof in both cases. We shall use again the above notation, the existence of the line d being

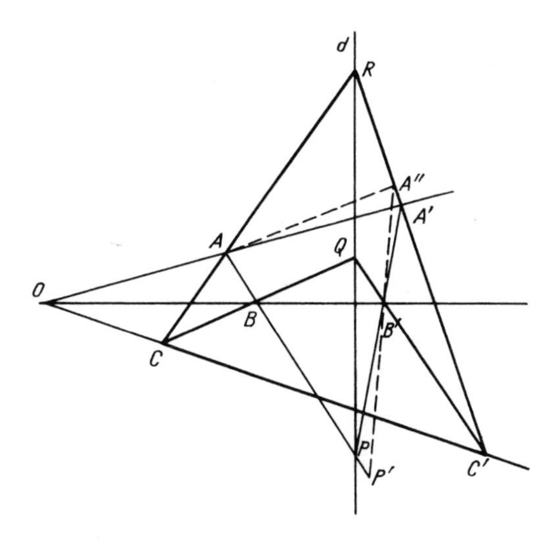

Fig. 9

assumed now. Suppose, for instance, $d \neq AB$, and denote $O = BB' \cap CC'$ (Fig. 9). Consider $A'' = OA \cap A'C'$. We may apply the direct part of the theorem to the triangles ABC, $A''B'C'$, which gives $P' = AB \cap A''B'' \in d$. But this implies $P' = P$, because, if not $AB = d$, contrary to our hypothesis. Then, we also have $A'' = A'$, q.e.d.

We shall need the following particular case of the previous theorem, expressed as a theorem, for the (nonenlarged) affine space.

13. THEOREM. Let ABC and $A'B'C'$ be two triangles in the usual affine space. Suppose that the lines AA', BB' and CC' are either concurrent or parallel and that $AB \parallel A'B'$, $AC \parallel A'C'$. Then, we also have $BC \parallel B'C'$. Conversely, if the corresponding sides of the two triangles are pairwise parallel, then the lines which join the corresponding vertices are either concurrent or parallel.

In fact, by the hypotheses, and using Theorem 12, we see that $AB \cap A'B'$, $AC \cap A'C'$ and $BC \cap B'C'$ are collinear points of the corresponding enlarged affine space, which lie on the same improper line (because the first two listed points are improper). Hence $BC \cap B'C'$ is an improper point, i.e., $BC \parallel B'C'$, q.e.d.

The converse also follows by Theorem 12, because its hypothesis implies the collinearity of $AB \cap A'B'$, $AC \cap A'C'$, $BC \cap B'C'$ in the enlarged space (actually, these points are on the improper line of the plane ABC).

An interesting fact related to Desargues' Theorem is that, in order to prove this theorem for coplanar triangles, we were obliged to use the three-dimensional space. The question is if this is due to the method of proof employed or if it is an intrinsic fact? In other words, we must determine if the Desargues plane theorem is dependent or not on the plane incidence and parallel axioms. As a matter or fact, we can show by a convenient model that the Desargues plane theorem is independent of the axioms mentioned.

Consider as points the usual points of a Euclidean plane (which can be defined axiomatically, but which we shall consider here intuitively, because we didn't yet state all the axioms of Euclidean geometry).

Fix an oriented axis a in this plane and let's call, arbitrarily, one of the two resulting half-planes *positive* and the other *negative*. As lines of the constructed geometry we shall define: the axis a, the lines parallel to a, the lines which enclose in the positive half-plane an angle $\alpha \geq \pi/2$ with the positive sense of a and, finally, the pairs (h,k) of half-lines both starting at a point of the axis a and which are such that h is in the positive half-plane and encloses an angle $\alpha < \pi/2$ with the positive sense of a, while k is in the negative half-plane and its extension in the positive half-plane encloses with the positive sense of the axis a an angle β such that $\operatorname{tg} \beta = 2 \operatorname{tg} \alpha$ (see Fig. 10).

Using both synthetic and analytic geometry of the Euclidean plane, it is possible to verify that these points and lines really satisfy Axioms I1, I2, I3, and II (exercise for the reader). Hence, we have an affine plane and a corresponding *enlarged or projective plane*.

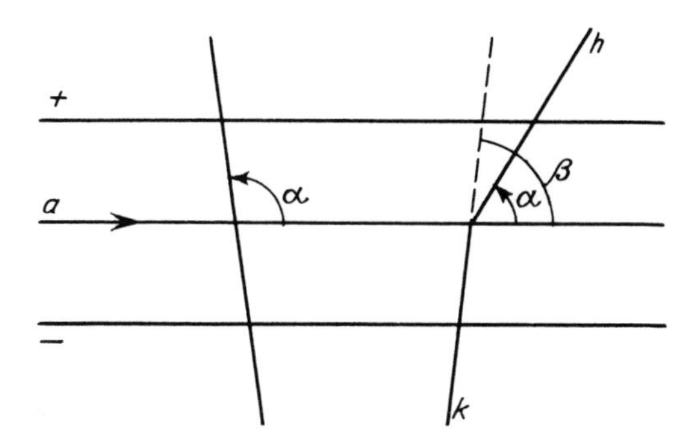

Fig. 10

It is easy to construct, in this plane, pairs of triangles satisfying the hypothesis but not the conclusion of the Desargues Theorem. Such are the triangles *ABC* and *A'B'C'* of Fig. 11 for the direct theorem and those of Fig. 12 for the converse. In Fig. 11, *AA'*, *BB'*

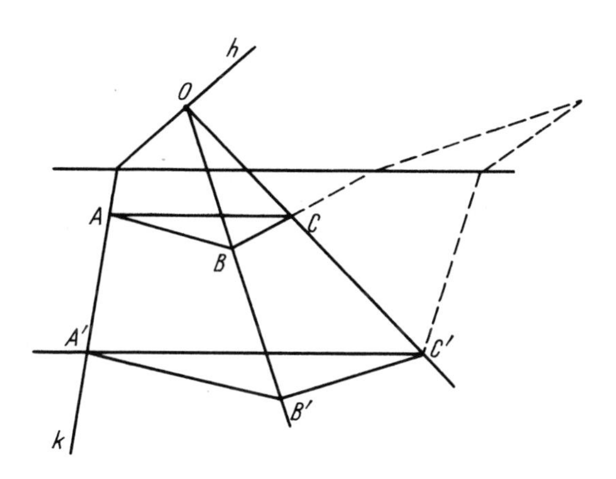

Fig. 11

and *CC'* are concurrent at *O*, in the sense of the model. *AA'* is a line of the type (h,k) above, $AC \parallel A'C'$, $AB \parallel A'B'$, but *BC* cannot be parallel with *B'C'* because this would contradict the converse of Theorem 13 in the Euclidean plane. In Fig. 12, the sides of the two triangles are pairwise parallel, which implies that *AA'*, *BB'*, *CC'* are concurrent in the Euclidean plane, but they are not concurrent

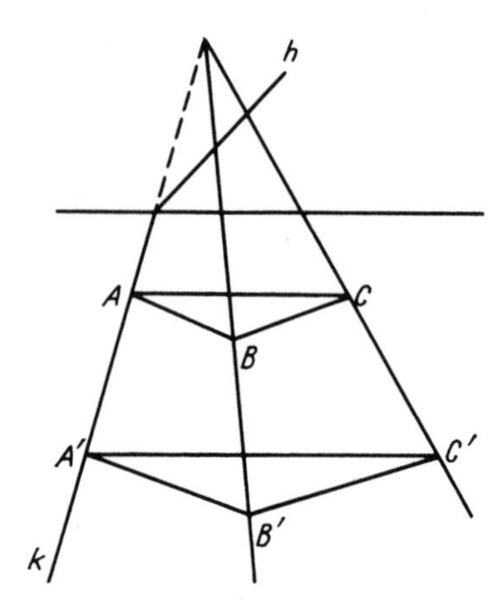

Fig. 12

in the sense of the model. As a matter of fact, we even proved the independence of the particular case of Theorem 13 of the axioms of plane affine geometry.

This model shows that there are affine planes, where Desargues' Theorem does not hold. These are called *non-Arguesian planes* (or *geometries*) and, clearly, such a plane cannot be embedded in three-dimensional space (affine or projective, according to the desired case), because in such a space we proved the Desargues theorem. In the remaining part of this book, we shall always consider planes which are embedded in a three-dimensional geometry, i.e., *Arguesian planes*.

We shall end this section with an application of the Desargues Theorem.

Generally, the notions and properties which may be defined and, respectively, proven, from the axioms of affine space only are called *affine notions and properties*. We want to define here an interesting affine notion, the *middle* of a segment. (In spite of the appearances, the middle of a segment may be defined without using distances.)

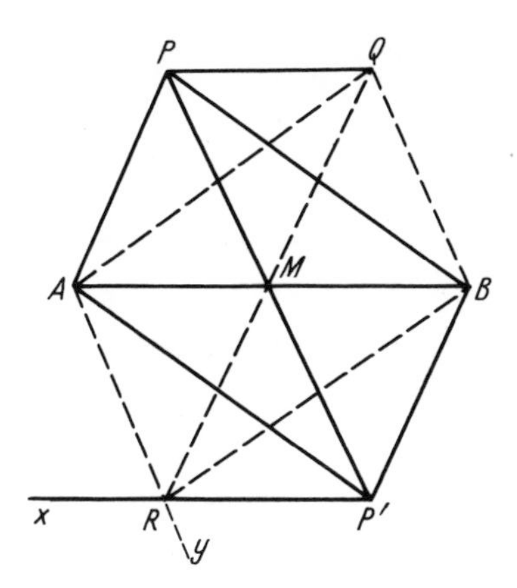

Fig. 13

Consider a segment AB (Fig. 13) and let P be a point not belonging to the line AB. Join P to A and B and consider a parallel to PB through A and a parallel to AP through B. These two lines meet at some point P'. Consider $M = AB \cap PP'$, *if such an* M *exists*. Then M is called the *middle* of the segment AB.

We may prove that M does not depend on the choice of P. In fact, let Q be another point and take $P'x \parallel PQ$, $Ay \parallel QB$ and $P'x \cap Ay = R$. The triangles BPQ and $AP'R$ have corresponding parallel sides and, by Theorem 13, PP', QR and AB are either concurrent or parallel. By applying next the direct part of the same Theorem 13 to the triangles APQ and $BP'R$, we get $AQ \parallel BR$, which shows that R plays the role of Q' (the point which plays with respect to Q the same role as P' to P). It follows that Q defines the same middle of AB as P, q.e.d.

The uniqueness of the middle point is thus proven. However, we didn't prove yet the existence of such a point and we shall consider this problem later.

§3. THE FUNDAMENTAL ALGEBRAIC STRUCTURES OF AN AFFINE SPACE

It is possible to attach to every affine space an abelian group defined by using some equivalence classes of pairs of points and a field defined by equivalence classes of triples of collinear points. Moreover, we shall see in the next section that these are related by the fact that the group has a vector space structure over the field. These algebraic structures are very important for affine space and we shall study them in this and the next section.

a. Vectors of an Affine Space

In order to get the group structure, we define first an *oriented segment*, which is to be an ordered pair of distinct points (A,B). Such a segment will be denoted by $\overrightarrow{AB}$, where A is called the *origin* and B the *end* of the segment. The line AB is called the *support* of the segment and two segments with the same support are called *collinear*, while two segments with coplanar supports are called *coplanar segments*.

We shall define now an equivalence relation on the set of the oriented segments.

1. DEFINITION. The oriented segments $\overrightarrow{AB}$, $\overrightarrow{A'B'}$ are called *strictly equipollent* if either $A = A'$ and $B = B'$ or $AB \parallel A'B'$ and $AA' \parallel BB'$.

This relation will be denoted by $\approx$ and two strictly equipollent segments either coincide or are noncollinear.

The relation $\approx$ is clearly reflexive and symmetrical, hence it may be extended to an equivalence relation (which must also be transitive) in the following way.

2. DEFINITION. Two oriented segments $\overrightarrow{AB}$ and $\overrightarrow{A'B'}$ are called *equipollent* if there is some finite sequence of distinct oriented segments $\overrightarrow{A_1B_1}, \ldots, \overrightarrow{A_nB_n}$ such that $\overrightarrow{AB} \approx \overrightarrow{A_1B_1}$, $\overrightarrow{A_1B_1} \approx$

$\overrightarrow{A_2B_2},\ldots,\overrightarrow{A_nB_n} \approx \overrightarrow{A'B'}$. In this case, we say that *we may go from $\overrightarrow{AB}$ to $\overrightarrow{A'B'}$ by n intermediary segments.*

This new relation will be denoted by $\overrightarrow{AB} \sim \overrightarrow{A'B'}$ and it is obviously an equivalence relation.

3. PROPOSITION. Any two noncollinear equipollent oriented segments are strictly equipollent.

We prove this by induction on the number n of intermediary segments. For $n = 0$ the result is trivial. Suppose it is true for all $n \leq m$ and let's prove it for $n = m + 1$. Suppose $\overrightarrow{AB} \sim \overrightarrow{A'B'}$ by

$$\overrightarrow{AB} \approx \overrightarrow{A_1B_1},\ \overrightarrow{A_1B_1} \approx \overrightarrow{A_2B_2},\ldots,\ \overrightarrow{A_{m+1}B_{m+1}} \approx \overrightarrow{A'B'}. \tag{1}$$

Suppose that there is some intermediary segment in (1) $\overrightarrow{A_iB_i}$ which is noncollinear with both $\overrightarrow{AB}$ and $\overrightarrow{A'B'}$. Then, by the induction hypothesis, we have $\overrightarrow{AB} \approx \overrightarrow{A_iB_i}$, $\overrightarrow{A_iB_i} \approx \overrightarrow{A'B'}$ (Fig. 14). Now, we may use Theorem 2.13 (Desargues) for the triangles AA_iA' and BB_iB' which gives $AA' \parallel BB'$. Since we also have $AB \parallel A'B'$, we get $\overrightarrow{AB} \approx \overrightarrow{A'B'}$.

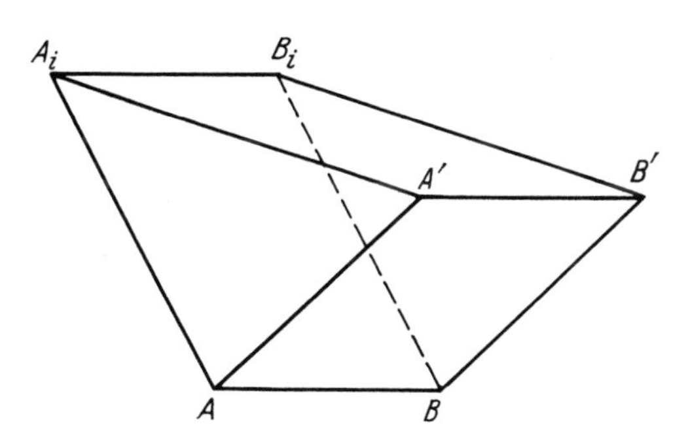

Fig. 14

Consider now the case where no $\overrightarrow{A_iB_i}$ as above exists. Then, all the intermediary segments of (1) are either on the line AB or on the

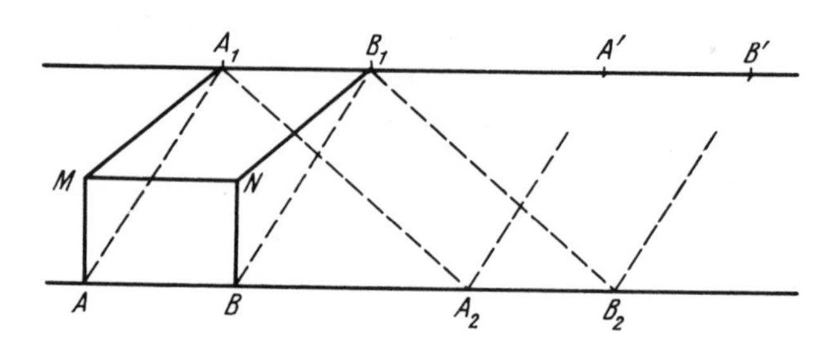

Fig. 15

line $A'B' \parallel AB$. Namely, those of an odd index are on $A'B'$ and those of an even index are on AB (Fig. 15).

Take a point M which is neither on AB nor on $A'B'$. For instance, we can take M outside the plane of the lines AB, $A'B'$. Let N be the intersection point of the parallel to AM through B with the parallel to A_1M through B_1. By using the converse part of Theorem 2.13 for the triangles AMA_1 and BNB_1, we deduce $\overrightarrow{MN} \approx \overrightarrow{A_1B_1}$ and $\overrightarrow{AB} \approx \overrightarrow{MN}$. Hence, by the already proven part of the Theorem, we get $\overrightarrow{MN} \approx \overrightarrow{A'B'}$, and then, like in the case of Fig. 14, $\overrightarrow{AB} \approx \overrightarrow{A'B'}$. Note that if the strong incidence axioms hold, M may be, for instance, the third point of AA_1 and N the intersection of BB_1 with the parallel to AB through M.

Note also that the result holds for plane Arguesian geometry even in the case where a point M as above does not exist (of course, we admit here only the usual incidence axioms). Then, in fact, the plane consists of points of AB and $A'B'$ only, and we must therefore have $AA' \parallel BB'$, which implies $\overrightarrow{AB} \approx \overrightarrow{A'B'}$, q.e.d. In this case, we also have $AB' \parallel BA'$, i.e., $\overrightarrow{AB} \approx \overrightarrow{B'A'}$ whence $\overrightarrow{AB} \approx \overrightarrow{BA}$.

Finally, it is also worth noting that the above proof does not hold in non-Arguesian planes.

4. COROLLARY. If $\overrightarrow{AB} \sim \overrightarrow{A'B'}$, it is always possible to go from the first segment to the second by at most one intermediary segment.

In fact, if the given segments are noncollinear we need no intermediary segments in view of Proposition 3. If they are collinear and we have the sequence (1), then $\overrightarrow{A_1B_1}$ is necessarily on a different line and, using again Proposition 3, it suffices to take $\overrightarrow{A_1B_1}$ as the unique intermediary segment.

For later algebraic requirements, it is useful to add to the set of oriented segments the pairs $(A,A) = \overrightarrow{AA}$, where the origin and end coincide. Then, we have to add the equipollences $\overrightarrow{AA} \sim \overrightarrow{BB}$ for every two points A,B, while $\overrightarrow{AA}$ is never equipollent to a segment with distinct ends. Thereafter, all the properties of oriented segments or equipollence must be considered for this case as well, and the reader is asked to carry out the corresponding proofs by himself. Clearly, the extended equipollence relation is also an equivalence relation.

5. DEFINITION. *A vector (free vector)* is a class of equipollent oriented segments.

We shall denote the vectors by small barred Latin letters, $\bar{u}$, $\bar{v}, \ldots$ and if we have to indicate also a representative segment of the class, we shall denote it as an oriented segment, $\overrightarrow{AB}$, but we shall speak of the *vector* $\overrightarrow{AB}$, not of the *segment* $\overrightarrow{AB}$. (The difference between a segment and its equipollence class is not expressed by the notation, but rather by replacing the word segment with the word vector).

The set of all vectors of the affine space will be denoted by V. This set has a distinguished element, defined by the class of all segments of the form $\overrightarrow{AA}$. This distinguished element is called the *zero vector* and is denoted by $\bar{0}$. Two vectors are called collinear if the segments of the respective classes are parallel. A vector $\bar{v}$ is said to belong to a plane α if the segments defining $\bar{v}$ are parallel to α.

The following result is very important.

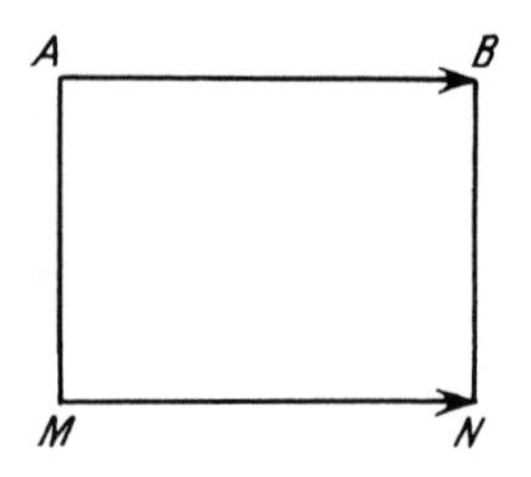

Fig. 16

6. THEOREM. For every point A and every vector $\bar{v}$, there is a unique point B that $\overrightarrow{AB} = \bar{v}$.

In fact, take $\overrightarrow{MN} = \bar{v}$ and suppose $A \bar{\in} MN$ (which is always possible). Then (Fig. 16), if we consider B at the intersection of the parallel to MN through A with the parallel to AM through N, we obviously have $\overrightarrow{AB} \sim \overrightarrow{MN}$, which proves the existence of the needed point. The uniqueness follows by reductio ad absurdum. If another point B' has the same property, the parallel axiom II is contradicted.

b. The Vector Sum

Now, we are able to get the first of the announced algebraic structures:

7. THEOREM. The set V of the vectors of an affine space has a natural structure of an additive abelian group.

To prove this, let $\bar{u}$, $\bar{v}$ be two vectors. Choose an arbitrary point A and, by applying Theorem 6 twice, construct the points B and C such that $\bar{u} = \overrightarrow{AB}$, $\bar{v} = \overrightarrow{BC}$. Then define

$$\bar{u} + \bar{v} = \overrightarrow{AC}.$$

Of course, we must prove that the result does not depend on the choice of A. If we take some other point A', and if we suppose first

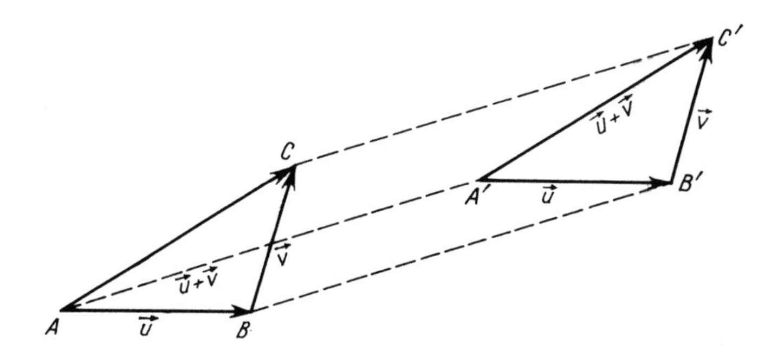

Fig. 17

that A' is neither collinear with A,B nor situated on the parallel through A to BC (Fig. 17), then, using the parallelism relations implied by the equipollences $\overrightarrow{AB} \sim \overrightarrow{A'B'}$ $(= \bar{u})$ and $\overrightarrow{BC} \sim \overrightarrow{B'C'}$ $(= \bar{v})$ and the Desargues Theorem for the triangles ABC, $A'B'C'$, we get $AC \parallel A'C'$. Hence, $\overrightarrow{AC} \sim \overrightarrow{A'C'}$ which proves our assertion. If A' does not satisfy the imposed conditions, we shall compare the sum constructed from both A and A' with the sum constructed from any point A'' which satisfies these conditions, and we get the same conclusion $\overrightarrow{AC} \sim \overrightarrow{A'C'}$, q.e.d.

Furthermore, we must verify the group properties.

The associativity follows trivially and we leave it to the reader. Equally trivial is the fact that the vector $\bar{0}$ is the unit of the sum, i.e.,

$$\bar{v} + \bar{0} = \bar{0} + \bar{v} = \bar{v}$$

for every vector $\bar{v}$.

Next, take any vector $\bar{v}$, consider $\bar{v} = \overrightarrow{AB}$ and define the *opposite vector* $-\bar{v} = \overrightarrow{BA}$, which does not depend on the choice of A as we can see immediately from the definition of the equipollence of oriented segments. We get

$$\bar{v} + (-\bar{v}) = (-\bar{v}) + \bar{v} = \bar{0} .$$

As usual, one writes $\bar{u} - \bar{v}$ instead of $\bar{u} + (-\bar{v})$.

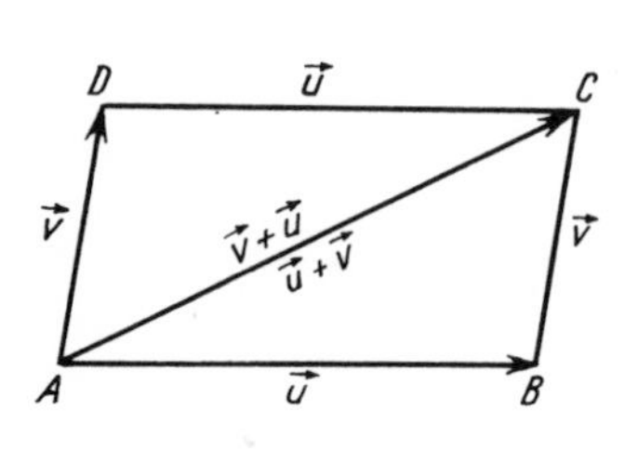

Fig. 18

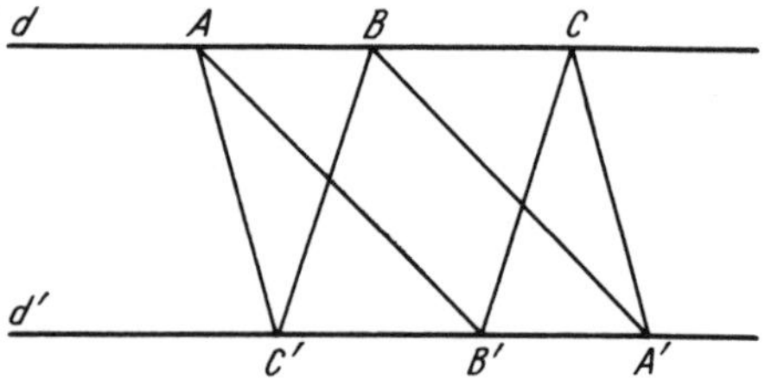

Fig. 19

It follows that V is an additive group and we have only to show now that the sum of vectors is commutative.

Consider first two noncollinear vectors $\bar{u}$ and $\bar{v}$. Put $\bar{u} = \overrightarrow{AB}$, $\bar{v} = \overrightarrow{BC}$ and, also, $\bar{v} = \overrightarrow{AD}$ (Fig. 18). Then, the parallelism relations implied by $\overrightarrow{AD} \sim \overrightarrow{BC}$ yield $\overrightarrow{DC} \sim \overrightarrow{AB}$ and we have

$$\bar{v} + \bar{u} = \overrightarrow{AC} = \bar{u} + \bar{v}.$$

Now, suppose that $\bar{u}$, $\bar{v}$ are two collinear vectors and take a vector $\bar{w}$ which is not collinear with them. Using the commutativity relation for noncollinear vectors we have

$$\begin{aligned}\bar{u} + \bar{v} &= (\bar{u} + \bar{v}) + (\bar{w} - \bar{w}) = [\bar{u} + (\bar{v} + \bar{w})] - \bar{w} \\ &= [(\bar{v} + \bar{w}) + \bar{u}] - \bar{w} = [\bar{v} + (\bar{w} + \bar{u})] - \bar{w} \\ &= [\bar{v} + (\bar{u} + \bar{w})] - \bar{w} = (\bar{v} + \bar{u}) + (\bar{w} - \bar{w}) = \bar{v} + \bar{u}\end{aligned}$$

which is just the desired relation.

8. COROLLARY. Let d, d' be two parallel lines, and $A,B,C \in d$, $A',B',C' \in d'$. Then, if $AB' \parallel BA'$ and $BC' \parallel CB'$, we also have $AC' \parallel A'C$.

In fact, if we denote $\overrightarrow{AB} = \bar{u}$, $\overrightarrow{BC} = \bar{v}$ (Fig. 19), we also have (in view of the hypotheses) $\overrightarrow{C'B'} = \bar{v}$, $\overrightarrow{B'A'} = \bar{u}$. Hence, $\overrightarrow{AC} = \bar{u} + \bar{v} = \bar{v} + \bar{u} = \overrightarrow{C'A'}$. It follows that $\overrightarrow{AC} \sim \overrightarrow{C'A'}$, which implies $AC' \parallel A'C$, q.e.d.

The proposition of Corollary 8 is a particular case of the so-called *Pappus-Pascal Theorem*. We shall see later on that the general case of this Theorem, when d and d' are concurrent, cannot be proven from the Axioms I and II alone.

9. COROLLARY. (The parallelogram property). If $\overrightarrow{AB} \sim \overrightarrow{A'B'}$, then we also have $\overrightarrow{AA'} \sim \overrightarrow{BB'}$.

In fact, if $\overrightarrow{AB}$ and $\overrightarrow{A'B'}$ are noncollinear the result is trivial. If they are collinear segments (Fig. 20), we have

$$\overrightarrow{AA'} = \overrightarrow{AB} + \overrightarrow{BA'}, \qquad \overrightarrow{BB'} = \overrightarrow{BA'} + \overrightarrow{A'B'} \quad ,$$

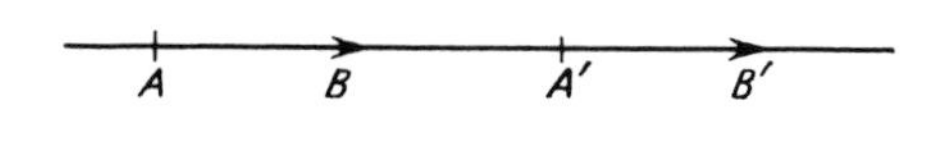

Fig. 20

and the result follows from the commutativity of the sum of two vectors.

c. Scalars of an Affine Space

In order to get the second algebraic structure asserted, we first fix a point O in the space and call it the *origin*. Then, if (M,N) is an ordered pair of points which are collinear with O, and $N \neq O$, it will be called an *O-pair* and denoted by $(M,N)_O$ or, if no confusion is possible, by (M,N).

10. DEFINITION. Two O-pairs (M,N) and (M',N') are called *strictly similar* if one of the following situations hold: a) $M = O = M'$; b) $M = N$ and $M' = N'$; c) $M = M'$ and $N = N'$; d) $MM' \parallel NN'$. This relation will be denoted by $(M,N) \approx (M',N')$. Two O-pairs (P,Q) and (P',Q') are called *similar* if it is possible to go from the first to the second by n distinct intermediary pairs:

$$(P,Q) \approx (P_1,Q_1),\ldots,(P_n,Q_n) \approx (P',Q') \tag{2}$$

In this case, we shall write $(P,Q) \sim (P',Q')$.

Note that if $(M,N) \approx (M',N')$ and these pairs correspond to case d) of the definition, they belong to distinct lines.

The relation $\approx$ is reflexive and symmetric and the relation $\sim$ is also transitive, i.e., this is an equivalence relation. Its equivalence classes will be called *scalars* and will be denoted by small Latin letters (but, of course, they are not to be confused with the lines of the space). Also, we shall denote scalars by arbitrary representative pairs of the corresponding classes. The set of all the scalars will be denoted by K_O or simply by K. Like for the equipollence relation, we shall prove

11. PROPOSITION. If $(P,Q) \sim (P',Q')$ and if these pairs belong to distinct lines, then $(P,Q) \approx (P',Q')$.

Of course, we may suppose $P \neq O \neq P'$ because, in the contrary case, the result follows from the definitions.

Then, Proposition 11 will be proven by induction on the number m of intermediary pairs. For $m = 0$, the result clearly holds. Suppose it holds for all $m \leq n - 1$ and suppose we have the relation (2). If there is an intermediary pair (P_i, Q_i) which is collinear with neither (P,Q) nor (P',Q'), then, by the inductive hypothesis, $(P,Q) \approx (P_i, Q_i) \approx (P',Q')$ (Fig. 21). This implies $PP_i \parallel QQ_i$, $P_iP' \parallel Q_iQ'$ and, by the Desargues Theorem, $PP' \parallel QQ'$ which means $(P,Q) \approx (P',Q')$.

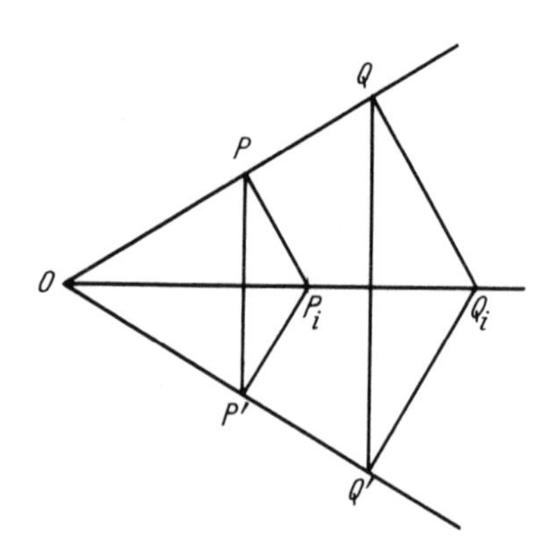

Fig. 21

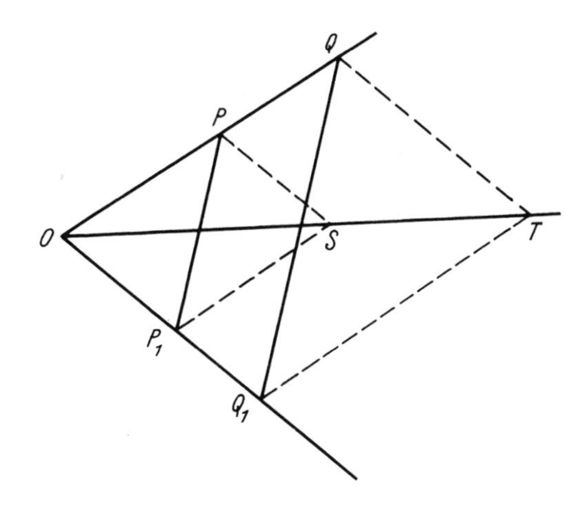

Fig. 22

If such an intermediary pair does not exist, we shall insert one between (P,Q) and (P_1,Q_1) in the following way (Fig. 22): take $PS \parallel OP_1$ and $P_1S \parallel OP$, join O to S and take T to be the intersection of OS with the parallel to PS through Q. This gives $(S,T) \approx (P,Q)$ and, by using again the Desargues Theorem, $(S,T) \approx (P_1,Q_1)$. Hence, (S,T) is the required intermediary pair and the proof of Proposition 11 will be completed in just the same way as for Proposition 3.

Like for Corollary 4, we have also

12. COROLLARY. If $(P,Q) \sim (P',Q')$ then we can go from one pair to the other either directly or by a single intermediary pair.

Another important result is

13. THEOREM. For every O-pair (P,Q) $(P \neq O)$ and for every point $P' \neq O$ there is a unique point Q' such that $(P,Q) \sim (P',Q')$. For every pair (P,Q) and every $Q' \neq O$ there is a unique point P' such that $(P,Q) \sim (P',Q')$.

In fact, if P' is not collinear with O,P then Q' is the intersection point of OP' with the parallel through Q to PP' (Fig. 23).

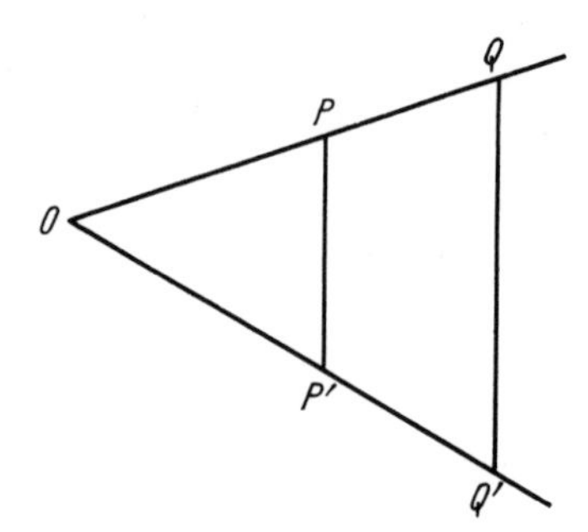

Fig. 23

If P' belongs to the line OP, we shall take first a point P'' not on this line and Q'' such that $(P,Q) \sim (P'',Q'')$, and, next, we construct Q' such that $(P'',Q'') \sim (P',Q')$. But then, $(P,Q) \sim (P',Q')$ and we have the required point (Fig. 24). The uniqueness follows by reductio ad absurdum.

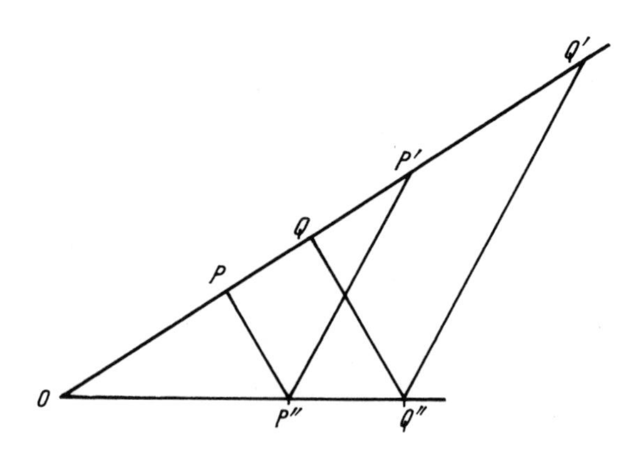

Fig. 24

Note also that when $P = O$, we must take necessarily $P' = O$ and Q' is indeterminate.

The second part of the theorem follows similarly by changing only the role of P and Q.

d. Scalar Algebraic Operations

Now, we shall define operations with scalars and prove

14. THEOREM. The set K may be endowed with the structure of a field.

We begin by defining the sum of two scalars.

Let $k_i \in K$ $(i = 1,2)$. By choosing an arbitrary point $Q \neq O$ and using the second part of Theorem 13, we may write in a unique manner $k_i = (P_i, Q)$. Then, let P be the point on OK such that

$$\overrightarrow{OP} = \overrightarrow{OP_1} + \overrightarrow{OP_2},$$

and put $k = (P,Q)$. We shall define

$$k = k_1 + k_2.$$

Let us show that the sum does not depend on the choice of the point Q. In fact, if we take another point Q' and if $Q' \bar{\in} OQ$ (Fig. 25), then P_1, P_2 will be replaced by the points P_1', P_2', such that $P_1P_1' \parallel QQ' \parallel P_2P_2'$, and P will be replaced by P', such that $\overrightarrow{OP'} = \overrightarrow{OP_1'} + \overrightarrow{OP_2'}$. It follows

$$\begin{aligned}\overrightarrow{PP'} &= \overrightarrow{OP'} - \overrightarrow{OP} = (\overrightarrow{OP_1'} + \overrightarrow{OP_2'}) - (\overrightarrow{OP_1} + \overrightarrow{OP_2}) \\ &= (\overrightarrow{OP_1'} - \overrightarrow{OP_1}) + (\overrightarrow{OP_2'} - \overrightarrow{OP_2}) = \overrightarrow{P_1P_1'} + \overrightarrow{P_2P_2'} \parallel \overrightarrow{QQ'},\end{aligned}$$

which shows that we get the same scalar k.

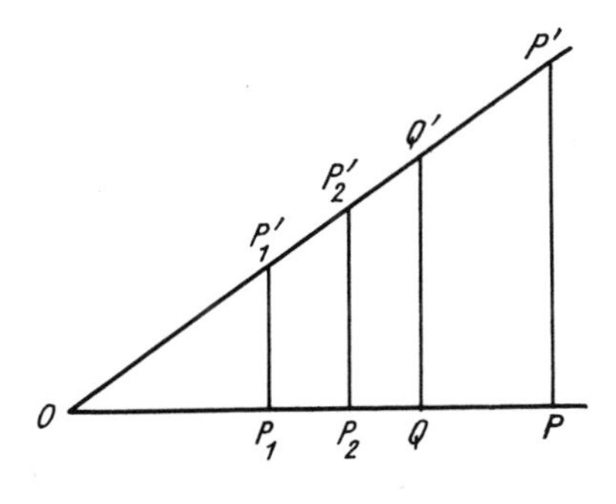

Fig. 25

If $Q' \in OQ$ we compare the respective scalars by using a third, intermediary point Q'' conveniently chosen and the conclusion is the same: the definition of the sum of scalars is correct.

The associativity and commutativity properties of the sum as defined above follows in a straightforward manner from the definition and the corresponding properties of the sum of vectors.

Next, from Definition 10, it follows that all the pairs (O,Q) with arbitrary Q are equivalent and define a scalar. We denote it by 0 and call it the *scalar zero*. It is trivial to verify that for every $k \in K$ we have $k + 0 = 0 + k = k$, which shows that 0 is the unit of the scalar sum.

Finally, take $k \in K$ and put $k = (P,Q)$. Consider the point P' defined by $\overrightarrow{OP'} = \overrightarrow{PO}$ and put $-k = (P',Q)$. This defines the *opposite* of because clearly $k + (-k) = (-k) + k = 0$. On the other hand, the reader will easily verify that $-k$ does not depend on the representation $k = (P,Q)$.

Hence K is an additive abelian group.

Let's go over now to the multiplication of scalars.

Take again k_1 and k_2 and choose $Q \neq O$. Next, put $k_2 = (P,Q)$ and suppose $k_2 \neq 0$. We have then $P \neq O$ which enables us to put $k_1 = (R,P)$. Now, we shall define $k_1 k_2 = (R,Q)$, and, if $k_2 = 0$, we define $k_1 k_2 = 0$. The order of these operations is essential. The proof of the independence of the product on the choice of Q is easy and we leave it to the reader.

The associativity of the product follows by the relations

$$[(S,R)(R,P)](P,Q) = (S,Q) \text{ or } 0,$$

$$(S,R)[(R,P)(P,Q)] = (S,Q) \text{ or } 0.$$

Furthermore, note that the scalar (Q,Q) does not depend on the choice of Q and

$$(Q,R)(R,R) = (Q,R) = (Q,Q)(Q,R).$$

Hence, if we put $1 = (Q,Q)$, this is the unit of the defined multiplication.

Consider the nonzero scalar $k = (P,Q)$ $(P \neq O)$. It has the associated scalar $k^{-1} = (Q,P)$, which is clearly independent on the choice of Q. We also see that $kk^{-1} = k^{-1}k = 1$, which proves the existence of the inverse for every scalar $k \neq 0$.

Now, in order to complete the proof of Theorem 14, we have to prove the distributivity relations

a) $(k_1 + k_2)k = k_1 k + k_2 k$,

b) $k(k_1 + k_2) = kk_1 + kk_2$.

For a) it suffices to take $k \neq 0$ (if $k = 0$, the result is trivial) and to put $k = (Q,P)$, $k_1 = (P_1,Q)$, $k_2 = (P_2,Q)$. Then we have

$$k_1 + k_2 = (\text{End of } (\overrightarrow{OP_1} + \overrightarrow{OP_2}),Q),$$

$$(k_1 + k_2)k = (\text{End of } (\overrightarrow{OP_1} + \overrightarrow{OP_2}),P),$$

$$k_1 k = (P_1,P),\ k_2 k = (P_2,P),$$

$$k_1 k + k_2 k = (\text{End of } (\overrightarrow{OP_1} + \overrightarrow{OP_2}),P),$$

which proves the desired equality.

The proof of b) is more complicated. Consider $k_1 \neq 0$, $k_2 \neq 0$, $k_1 + k_2 \neq 0$ and $k_1 = (P_1,Q)$, $k_2 = (P_2,Q)$, $k_1 + k_2 = (P,Q)$, $k = (S_1,P_1) = (S_2,P_2) = (S,P)$, $kk_1 + kk_2 = (T,Q)$, $k(k_1 + k_2) = (S,Q)$ (Fig. 26).

In order to show the desired equality, we have to show $S = T$.

To do this take $(B,A) = k$ on another line through O. We have then $AP_1 \parallel BS_1$, $AP_2 \parallel BS_2$ and $AP \parallel BS$, from which we shall derive the result if we prove $BT \parallel AP$. Take $\overrightarrow{AX} = \overrightarrow{OP_1}$ and $\overrightarrow{BY} = \overrightarrow{OS_1}$. The triangles AXP_1 and BYS_1 have parallel corresponding sides, whence, by the converse Desargues Theorem, the points O, X and Y are collinear. Further, from $\overrightarrow{OP_1} + \overrightarrow{OP_2} = \overrightarrow{OP}$ it follows easily $\overrightarrow{OP_1} = \overrightarrow{P_2P} = \overrightarrow{AX}$, which implies $XP \parallel AP_2$ and, similarly, from $\overrightarrow{OS_1} + \overrightarrow{OS_2} = \overrightarrow{OT}$ it follows $YT \parallel BS_2 \parallel XP$.

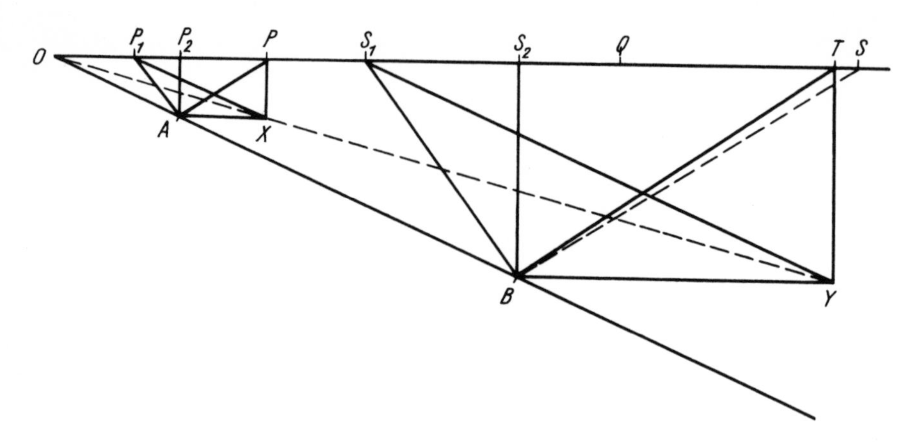

Fig. 26

Finally, by applying the Desargues Theorem to the triangles *AXP* and *BYT*, we get $AP \parallel BT$ which completes the proof.

The reader will analyze by himself the cases where one of the scalars $k_1, k_2, k_1 + k_2$ vanishes.

e. Properties of the Scalar Field

Let us also make a few other considerations about the field K. First, we have

15. PROPOSITION. By changing the origin O, one gets an isomorphic scalar field.

In fact, take a new origin O' and consider the correspondence $(P,Q)_O \to (P',Q')_{O'}$, defined by the conditions $\overrightarrow{OP} = \overrightarrow{O'P'}$, $\overrightarrow{OQ} = \overrightarrow{O'Q'}$. This is a one-to-one correspondence and sends similar O-pairs to similar O'-pairs. Actually, if $(P,Q)_O \approx (P_1, Q_1)_O$, and if we suppose that we are in the case d) of Definition 10 (which is the only one of interest because all the others are trivial), we shall also have $(P_1, Q_1)_O \to (P_1', Q_1')_{O'}$ with $\overrightarrow{OP_1} = \overrightarrow{O'P_1'}$ and $\overrightarrow{OQ_1} = \overrightarrow{O'Q_1'}$ (Fig. 27). This implies $\overrightarrow{PP'} = \overrightarrow{QQ'} = \overrightarrow{P_1P_1'} = \overrightarrow{Q_1Q_1'} (= \overrightarrow{OO'})$, whence we derive $P'P_1' \parallel PP_1 \parallel QQ_1 \parallel Q'Q_1'$, which proves our assertion.

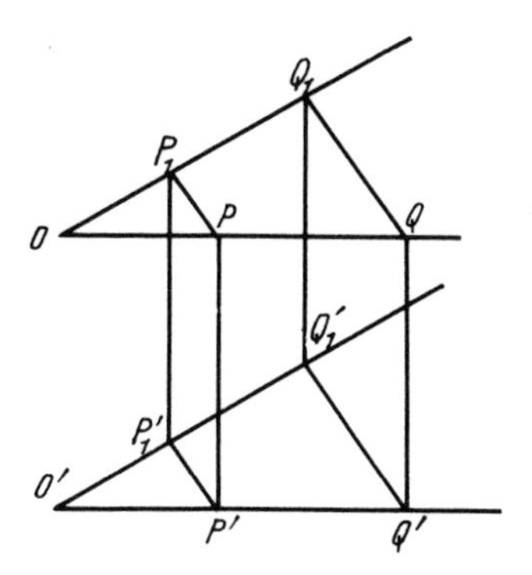

Fig. 27

Hence, if we take $K' = K_{O'}$ the previous correspondence induces a map $f : K \to K'$, and the reader will easily prove that f is an isomorphism, q.e.d.

It follows that the scalar field is defined by the affine space up to an isomorphism which, of course, is not essential. Thereafter, when speaking about the scalar field K, we always suppose that an origin was chosen but it is unimportant which concrete point it is.

It is of interest to note that

16. REMARK. The field K has the same (cardinal) number of points as the set of the points on a line.

In fact, by the very definition of K, we get a one-to-one correspondence between the elements of K and the set of the points of the line OQ.

Implicitly, all the lines have the same (cardinal) number of points. (We leave to the reader to prove this in a straightforward manner.) Furthermore, *the strong incidence axioms hold iff* K *has at least three elements.*

17. PROPOSITION. The scalar field K is of characteristic 2 iff there is a parallelogram with parallel diagonals.

Of course, by a parallelogram we understand here a quadruple of points $ABCD$ such that $\overrightarrow{AB} = \overrightarrow{DC}$, and the terms *vertex, side* and *diagonal* have the usual meaning.

Now, if $AC \parallel BD$, we get $\overrightarrow{AB} = \overrightarrow{CD}$, whence $\overrightarrow{CD} = \overrightarrow{DC}$, $2\overrightarrow{CD} = \bar{0}$. By choosing the origin O, and $\overrightarrow{OP} = \overrightarrow{CD}$, the previous relation means $2k = 0$ for every scalar $k = (P,Q)$ and K is of characteristic 2.

Going back we see that, conversely, $2k = 0$, implies $AC \parallel BD$.

Proposition 17 is thus proved and we see that, moreover, every parallelogram has parallel diagonals.

The fact that the scalar field has or has not characteristic 2 is also related to the existence of the middle of a segment. First, let us prove

18. PROPOSITION. The point M of the line AB is the middle of the segment AB iff the ordered segments $\overrightarrow{AM}$ and $\overrightarrow{MB}$ are equipollent.

If $\overrightarrow{AM} \sim \overrightarrow{MB}$, there is some segment $\overrightarrow{PQ}$ on another line, which is equipollent to them (Fig. 28). Join P with B and take $P'A \parallel PB$, $P'B \parallel PA$. Using the parallelism relations implied by the equipollences considered and the Desargues Theorem for AMP' and PQB, we get $QB \parallel MP'$.

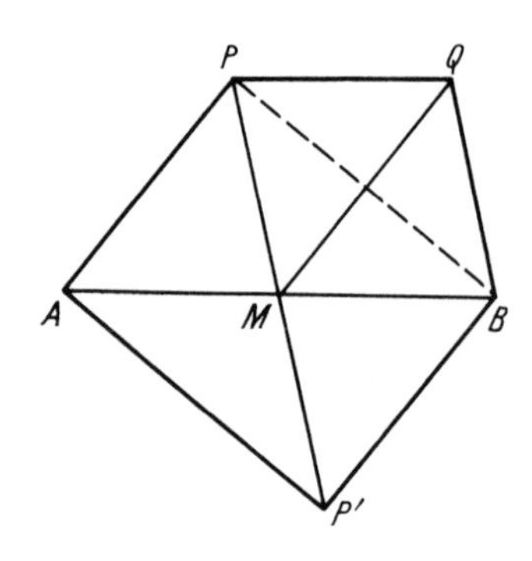

Fig. 28

Hence, P, M and P' are collinear and M is the middle of AB as defined at the end of §2.

Conversely, if M is the middle of AB and we consider some $\overrightarrow{PQ} \sim \overrightarrow{AM}$, we may use again the Desargues Theorem as above and derive from $QB \parallel MP'$ that $QB \parallel MP$. This implies $\overrightarrow{PQ} \sim \overrightarrow{MB}$, whence $\overrightarrow{AM} \sim \overrightarrow{MB}$ as announced in the proposition.

Using Proposition 18, we can establish now

19. THEOREM. If the field K is not of characteristic 2, every segment has a midpoint. Conversely, if at least one segment has a middle, all segments do, and the scalar field K has the characteristic $\neq 2$.

In fact, suppose K has characteristic $\neq 2$. Then the scalar $\frac{1}{2}$ exists and, if we take any segment AB, we may represent this scalar starting from the origin A. I.e., by considering $1 = (B,B)_A$, some point M exists such that $(M,B)_A = \frac{1}{2}$. Then, because $\frac{1}{2} + \frac{1}{2} = 2 \cdot (\frac{1}{2}) = 1$, we obviously have $\overrightarrow{AM} \sim \overrightarrow{MB}$ and, by Proposition 18, M is the middle of AB.

Conversely, if the segment AB has the middle M, we have $\overrightarrow{AM} = \overrightarrow{MB}$, whence $(M,B)_A = \frac{1}{2}$. It follows that $2 \neq 0$ and the characteristic of K is not 2.

Now, every segment has a midpoint because of the direct part of our theorem.

There is one more important problem which we want to study, namely, when is the scalar field commutative?

In order to answer this question, take the points P,Q,R and P',Q',R' (Fig. 29), on two different lines through the origin O, such that $(R',Q') \sim (Q,R)$ and $(Q',P') \sim (P,Q)$. In other words, $P'Q \parallel Q'P$ and $Q'R \parallel R'Q$. Then, we have

$$(P,Q) \cdot (Q,R) = (P,R) \quad , \quad (R',Q') \cdot (Q',P') = (R',P').$$

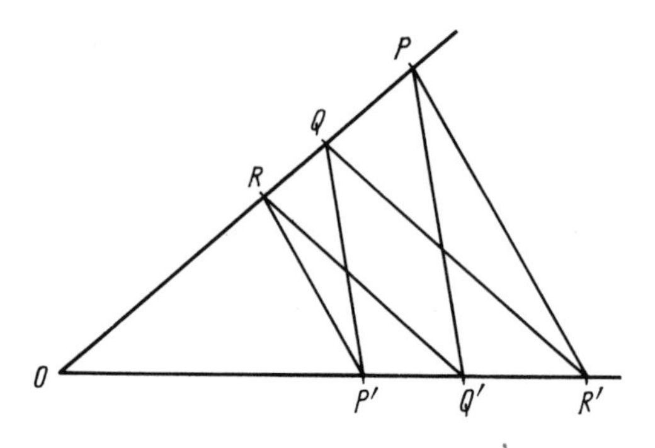

Fig. 29

The two considered products are just the two possible orders of multiplication of two scalars. The results are equal iff $(P,R) \sim (R',P')$, i.e., $PR' \parallel P'R$.

Consequently, we have

20. THEOREM. The scalar field K is commutative iff whenever we have, on two concurrent lines respectively, the triples of points P,Q,R and P',Q',R', then $P'Q \parallel Q'P$ and $Q'R \parallel R'Q$ imply $P'R \parallel R'P$.

The geometric property contained in Theorem 20 (and which we also discussed in relation with Corollary 8 is called the (*affine* or *special*) *Pappus-Pascal Theorem*. One can see that, as a matter of fact, this theorem comes from the more general projective property that: *for any configuration like the one of Fig. 29 in projective space, the points $P'Q \cap Q'P$, $Q'R \cap R'Q$ and $P'R \cap R'P$ are collinear.* However, we shall not study this general property here. Of course, in the case of Theorem 20, the three points considered are on the same improper line.

Later on, we shall see that the Pappus-Pascal Theorem is independent of the axioms of affine space and that it is also independent of the plane affine axioms and the Desargues Theorem. This means that there are also *non-Pappian geometries*, i.e., geometries where the Pappus-Pascal Theorem is not true.

§4. COORDINATES IN AFFINE SPACES

a. The Linear Space Structure

In this section, we shall study the relations between the two algebraic structures mentioned in the preceeding section, by defining an operation of a left multiplication of a vector with a scalar. In this operation, the scalars are to be considered in the following way: if an origin O is fixed, they are classes of O-pairs like in §3, and if we take another origin O', we identify classes of O-pairs with classes of O'-pairs which correspond to each other through the isomorphism defined in §3 between the corresponding fields K_O and $K_{O'}$.

Let $\bar{v}$ be a vector and k a scalar, and suppose we choose the origin O. Then, we define the vector $k\bar{v}$ as follows: if either $k = 0$ or $\bar{v} = 0$, $k\bar{v} = 0$; if $k \neq 0$, $\bar{v} \neq 0$, take $\bar{v} = \overrightarrow{OA}$, define B such that $k = (B,A)_O$ and put finally $k\bar{v} = \overrightarrow{OB}$. The independence of the vector obtained on the origin O follows immediately from the convention about identifying scalars. We leave the details to the reader. The same for

1. PROPOSITION. Two vectors are collinear iff they are proportional (i.e., one is obtained from the other by a multiplication with a scalar).

The properties of the operation defined are given by

2. THEOREM. The multiplication of a vector and a scalar makes V a left linear space over K.

In fact, we clearly have $1\bar{v} = \bar{v}$. If $k_1, k_2 \in K$ and $\bar{v} \in V$ we may consider

$$\bar{v} = \overrightarrow{OA}, \quad k_2 = (B,A), \quad k_2\bar{v} = \overrightarrow{OB}, \quad k_1 = (C,B),$$

$$k_1(k_2\bar{v}) = \overrightarrow{OC}, \quad (k_1k_2) = (C,B)(B,A) = (C,A),$$

$$(k_1k_2)\bar{v} = \overrightarrow{OC},$$

which proves $k_1(k_2\bar{v}) = (k_1k_2)\bar{v}$. Next, the distributivity relation

$$(k_1 + k_2)\bar{v} = k_1\bar{v} + k_2\bar{v}$$

is a straightforward consequence of the definitions, and the relation

$$k(\bar{v}_1 + \bar{v}_2) = k\bar{v}_1 + k\bar{v}_2$$

is represented by the same geometric configuration as the scalar relation $k(k_1 + k_2) = kk_1 + kk_2$, whence it is proved in just the same way as the last (§3).

This completes the proof of Theorem 2.

It follows that we may use for V the usual notions of linear algebra. For instance, Proposition 1 above means that: *two vectors are collinear iff they are linearly dependent.*

One also has

3. THEOREM. The linear space V is three-dimensional. The set of all vectors of a plane is a two-dimensional subspace of V and conversely.

In fact, by Axiom I7 we have four noncoplanar points O,A,B,C which define three vectors $\overrightarrow{OA}$, $\overrightarrow{OB}$, $\overrightarrow{OC}$. If they were linearly dependent we would have, for instance, $\overrightarrow{OC} = k_1\overrightarrow{OA} + k_2\overrightarrow{OB}$, which, in view of Proposition 1 and of the definition of the sum of vectors, would imply that C belongs to the plane OAB, and this is not true. Hence, V has three linearly independent vectors.

Next, consider a new vector $\bar{v} = \overrightarrow{OM}$ (Fig. 30). Denote by M' the intersection point of the parallel to OC through M with the plane OAB. Then, we clearly have $\overrightarrow{OM} = \overrightarrow{OM'} + \overrightarrow{M'M} = \overrightarrow{OM'} + \overrightarrow{OM_3}$, where $\overrightarrow{OM_3} \sim \overrightarrow{M'M}$. If, now, M_1 is the intersection of OA with the parallel through M' to OB, we have $\overrightarrow{OM'} = \overrightarrow{OM_1} + \overrightarrow{M_1M'} = \overrightarrow{OM_1} + \overrightarrow{OM_2}$, where $\overrightarrow{OM_2} \sim \overrightarrow{M_1M'}$. Thus,

$$\overrightarrow{OM} = \overrightarrow{OM_1} + \overrightarrow{OM_2} + \overrightarrow{OM_3}$$

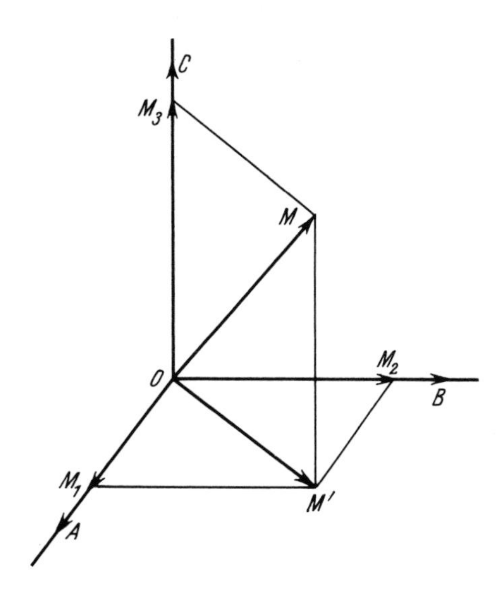

Fig. 30

and by using again Proposition 1

$$\overrightarrow{OM} = \overrightarrow{k_1 OA} + \overrightarrow{k_2 OB} + \overrightarrow{k_3 OC}.$$

Hence, every vector is a linear combination of three independent vectors, which proves the first assertion of Theorem 3. The second assertion is the same as the one proven above for the vector $\overrightarrow{OM'}$ in the plane OAB. Hence Theorem 3 is completely proven.

These results permit the consideration of bases and coordinates of the vectors of an affine space, which will be those of V.

A *basis* will be a system of three linearly independent vectors, usually denoted by $\bar{e}_i$ $(i = 1,2,3)$, and an arbitrary vector $\bar{v}$ has a unique representation of the form $\bar{v} = \sum_{i=1}^{3} x^i \bar{e}_i$ where $x^i \in K$. x^i will be the *coordinates* of the vector $\bar{v}$. If we also consider another basis $\bar{e}_i'$, we have some transformation formulas which are obviously of the form

$$\bar{e}_i' = \sum_{j=1}^{3} p_i^j \bar{e}_j \quad , \quad \bar{e}_i = \sum_{j=1}^{3} q_i^j \bar{e}_j', \tag{1}$$

where the matrix (q_i^j) is the inverse of the matrix (p_i^j), which means

$$\sum_{j=1}^{3} p_i^j q_j^k = \sum_{j=1}^{3} q_i^j p_j^k = \delta_i^k, \tag{2}$$

where δ_i^k are the so-called Kronecker symbols which are equal to 0 for $i \neq k$ and 1 for $i = k$.

With respect to the two bases, we have

$$\bar{v} = \sum_{i=1}^{3} x^i \bar{e}_i = \sum_{j=1}^{3} x'^j \bar{e}_j', \tag{3}$$

and using (1) we get the coordinate transformations

$$x'^j = \sum_{j=1}^{3} x^j q_j^i, \qquad x^i = \sum_{j=1}^{3} x'^j p_j^i. \tag{4}$$

b. Frames and Coordinates

Further, using the vector coordinates, we may define point coordinates and derive the so called *analytic geometry* of affine space.

To do this, let us remark first that, by fixing an origin O, the points M of the space will be in a one-to-one correspondence with the vectors $\overrightarrow{OM}$, called the *radius-vectors* with origin O and usually denoted by $\bar{r}_M$ or $\bar{r}$.

Next, by fixing a basis, we get a one-to-one correspondence between the points M and the coordinates of the radius-vector. The set $\{O, \bar{e}_1, \bar{e}_2, \bar{e}_3\}$ consisting of an origin O and a basis $\{\bar{e}_i\}$ of V $(i = 1,2,3)$ is called an *affine frame* in the space, and if we put

$$\bar{r}_M = \sum_{i=1}^{3} x^i \bar{e}_i, \tag{5}$$

the elements x of K are called the *coordinates* of the point M with respect to the given frame.

If we go to another origin O', we clearly have the following relation between the radius-vectors

$$\bar{r}' = \bar{r} - \overrightarrow{OO'} \tag{6}$$

and, next, considering coordinates by (5), and using the relation (1) for changing the basis, we get the relations between the coordinates of M with respect to two frames. These relations are

$$x'^i = \sum_{j=1}^{3} x^j q_j^i + a^i \quad , \quad x^i = \sum_{j=1}^{3} x'^j p_j^i + b^i , \tag{7}$$

where p and q are the matrices of (1), a^i are the coordinates of the vector $\overrightarrow{O'O}$ with respect to the basis $\bar{e}_i'$ and b^i are the coordinates of $\overrightarrow{OO'}$ with respect to the basis $\bar{e}_i$.

These are important formulas because the geometric properties are those which do not depend on the chosen frame, i.e., they are the properties invariant under the transformations (7).

The representation of points by coordinates is called the *analytic representation* and we intend now to get such representations for the lines and the planes of the space.

Choose a frame, and take some plane α (Fig. 31). Let M_0 be a fixed point and M an arbitrary point of α. Then, we have

$$\bar{r} = \bar{r}_0 + \overrightarrow{M_0M},$$

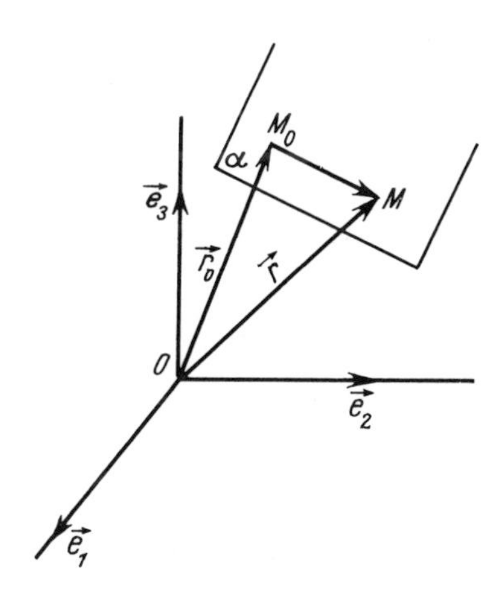

Fig. 31

where $\overrightarrow{M_0M}$ is a vector of α. But, in view of Theorem 3, the vectors $\overrightarrow{M_0M}$ form a two-dimensional subspace of V, whence, if we fix two non-collinear vectors $\bar{u}$, $\bar{v}$ in α, we see that a point M belongs to α iff its radius-vector is expressible as

$$\bar{r} = \bar{r}_0 + k_1\bar{u} + k_2\bar{v}. \tag{8}$$

Because of this fact, we say that (8) is the *parametric equation* of the plane α and k_1, k_2 are parameters.

Considering the coordinate expression of (8), we have (in an obvious notation)

$$\begin{aligned} x^1 &= x_0^1 + k_1u^1 + k_2v^1, \\ x^2 &= x_0^2 + k_1u^2 + k_2v^2, \\ x^3 &= x_0^3 + k_1u^3 + k_2v^3. \end{aligned} \tag{9}$$

Because $\bar{u}$ and $\bar{v}$ are linearly independent $\bar{v} \neq \bar{0}$, and suppose, for instance, $v^1 \neq 0$. Then the first equation (9) yields

$$k_2 = x^1(v^1)^{-1} - x_0^1(v^1)^{-1} - k_1u^1(v^1)^{-1},$$

which may be replaced in the other two equations (9). After a purely technical computation, which we leave to the reader, we see then that k_1 may be eliminated in a similar manner (again, the independence, i.e., nonproportionality, of $\bar{u}$ and $\bar{v}$ will be used). Hence (9) may be replaced by a single equation of the form

$$\sum_{i=1}^{3} x^i a_i + a = 0, \tag{10}$$

where not all a_i vanish. Let us mention that we didn't use the classical parameter elimination because, generally, the field K is not commutative.

Hence, the points of the plane α are satisfying the equation (10). Conversely, if the coordinates of a point are satisfying (10), the system (9) has solutions for k_1 and k_2, and that point belongs to the plane α. It follows that α is the locus of all the points satisying (10).

Now, let us start from an equation (10) with at least one non-vanishing coefficient a_i. Suppose, for instance, $a_1 \neq 0$ and multiply by a_1^{-1}. Then we obtain

$$x^1 = x^2 b + x^3 c + d,$$

which, together with $x^2 = x^2$, $x^3 = x^3$ and compared with (9) shows that the geometric locus of the points whose coordinates satisfy the given equation is the plane passing through the point $(d,0,0)$ and the vectors $(b,1,0)$ and $(c,0,1)$; x^2 and x^3 have the role of the parameters k_1 and k_2.

Hence, we proved

4. THEOREM. A plane of affine space is the locus of the points whose coordinates satisfy a linear equation of the form (10), where at least one of the coefficients a_i is non zero.

Obviously, the form of the equation (10) is unchanged by transformations (7).

If

$$\sum_{i=1}^{3} x^i a_i + a = 0, \qquad \sum_{i=1}^{3} x^i b_i + b = 0 \tag{11}$$

are the equations of two planes, the common points of these planes are given by the solutions of the system (11) and they must be found by the "substitution method" without changing the order of the products because K may be a non-commutative field.

As an exercise, the reader will prove that *the two planes are parallel (i.e., the system (11) is noncompatible) iff the coefficient* a_i *and* b_i *(but not* a *and* b*) are proportional.* If this does not happen, (11) are the *equations of the intersection line* of the two planes. Since every line can be obtained in this way, we proved

5. THEOREM. A line is a locus of points whose coordinates satisfy a system of linear equations of the form (11), where the coefficients a_i, b_i are not proportional.

Another analytic representation of a line, by the so-called *parametric equations*, can be obtained if we define geometrically the line by a point and by a *direction vector* (which is, simply, a vector parallel to that line).

Using Proposition 1, and acting as in case of the equation (8), we get that the line consists of all those points whose radius-vectors satisfy the relation

$$\bar{r} = \bar{r}_0 + t\bar{v}, \tag{12}$$

where $\bar{v}$ is the mentioned direction vector.

The coordinates of a direction vector of a line are called the *direction coefficients* of that line and these are three elements of K not all zero (because $\bar{v}$ above is non zero) and defined up to left proportionality. It is then obvious that two lines are parallel iff they have left proportional direction coefficients.

The previous remark indicates how to introduce coordinates in the enlarged affine space. Actually, from this remark it follows that the improper points are in a one-to-one correspondence with the systems of direction coefficients, i.e., by a trick whose motivation will be immediately clarified, with the systems $(\rho u^1, \rho u^2, \rho u^3, 0)$, where $\rho, u^i \in K$, $\rho \neq 0$ is arbitrary and not all of the u^i $(i = 1,2,3)$ vanish.

As for the proper points, we shall define the *homogeneous coordinates* of the point. Namely, if x^1, x^2, x^3 are the usual coordinates of the point, we introduce y^a $(\alpha = 1,2,3,4)$ by $y^i = \rho x^i$, $y^4 = \rho.1$ $(i = 1,2,3)$. These are four elements of K not all zero, and defined up to an arbitrary non zero left proportionality factor ρ. Clearly, we have a one-to-one correspondence between the points and their homogeneous coordinates (y^a).

Collecting these results, we find that, with respect to an affine frame, the points of the enlarged space are representable by *homogeneous coordinates*, which are systems (ρy^a) of elements of K, defined up to an arbitrary non-zero left proportionality factor and which are not all zero. Then, the improper points are characterized by $y^4 = 0$ and we can say that this is the equation of the improper plane of the space.

For a proper plane of the enlarged space, we have an equation of the form (10), and there we can replace $x^i = \rho^{-1} y^i = (y^4)^{-1} y^i$, which gives an equation of the form

$$\sum_{a=1}^{4} y^a a_a = 0. \tag{13}$$

This will be the equation of the plane in homogeneous coordinates. It is clearly a linear homogeneous equation with respect to the y^a, and it is easy to see that it is also satisfied by the improper points of the plane.

Since $y^4 = 0$ is a similar kind of equation, it follows that every plane of the enlarged space is represented analytically by an equation of the form (13), where at least one of the coefficients is non-zero.

It is simple to see that a line of the enlarged space will be represented by a system of two equations of the form (13) with non-proportional coefficients a_a.

If instead of doing three-dimensional geometry, we did plane Arguesian geometry, then, either directly or by supposing that we embedded our plane in a space and chose a frame such that our plane is $x^3 = 0$, we see that the points may be represented by pairs of non-homogeneous or triples of homogeneous coordinates. Also, the lines will be represented by one linear equation with respect to these coordinates. We have the well known situation of classical elementary analytic geometry with the only difference that the coefficients in the equation of a line are the right factors.

c. Affine Spaces over a Field

In connection with the present discussion, it is natural to ask whether any field K is associated with an affine geometry. We shall see that the answer is affirmative, i.e.,

6. THEOREM. If K is an arbitrary field, there are affine spaces for which K is the associated scalar field.

The proof of this theorem consists of constructing the required affine spaces and, from the developments above, we see how to do this.

Let us call a *point* every ordered triple (x^i) $(i = 1,2,3)$ of elements of K, a *line* every locus of the type given by Theorem 5, and a *plane* every locus as given by Theorem 4. Now we have to prove that these elements satisfy the axioms of affine geometry.

We begin by remarking, using a straightforward computation, that a system (11), which defines a line, is always equivalent to at least one system of the form

$$\begin{aligned} x^1 &= x^3 k + \ell, & x^2 &= x^1 p + q, & x^3 &= x^2 e + f, \\ x^2 &= x^3 h + m, & x^3 &= x^1 r + s, & x^1 &= x^2 g + t, \end{aligned} \tag{14}$$

which means that we may always look for the equations (14) of the required lines.

Let x^i_a $(i = 1,2,3;\ a = 1,2)$ be two distinct points and let, for instance, $x^3_1 \neq x^3_2$. Then, there is a line with equations of the first of the types (14), which contains them. In fact, the respective conditions are

$$\begin{aligned} x^1_1 &= x^3_1 k + \ell, & x^1_2 &= x^3_2 k + \ell, \\ x^2_1 &= x^3_1 h + m, & x^2_2 &= x^3_2 h + m, \end{aligned}$$

which imply

$$k = (x^3_1 - x^3_2)^{-1}(x^1_1 - x^1_2), \qquad h = (x^3_1 - x^3_2)^{-1}(x^2_1 - x^2_2)$$

and further, ℓ and m may be easily obtained.

Next, if we look at any other equations (14) such that the corresponding line contains the two given points, we find a system which is equivalent to the previous one.

We thus proved that Axiom I1 holds.

I2 holds because K has at least two elements which we can use as values of the parameter x^3, x^2 or x^1 in (14) and get two distinct points on every line. The strong form of Axiom I2 holds (i.e., we have I2') iff K has at least three elements.

For I3, we prove by a straightforward computation and using again (14) that the line passing through (0,0,0) and (1,0,0) does not contain the point (0,1,0).

In order to verify I4 (without using a general theory of homogeneous linear systems of equations over an arbitrary, generally noncommutative field), we shall note that equation (10) is equivalent to at least one of the following types of equations:

$$x^1 = x^2 a + x^3 b + c,\quad x^2 = x^3 a + x^1 b + c,\quad x^3 = x^1 a + x^2 b + c. \tag{15}$$

Next the plane containing three given points is obtained by a computation analogous to that which we used to find the line joining two points. This long but purely technical computation will be left to the reader.

It follows, in an obvious manner, that any plane has points.

Thus, I4 is satisfied.

I5 is proven by a straightforward computation, writing down the equations of the respective line. If the two points of Axiom I5 are (x_u^i) $(u = 1,2)$ we see that their line has the parametric equations

$$x^i = (1-h)x_1^i + hx_2^i$$

and, if (x_u^i) satisfy (10), i.e., we have $\sum_{i=1}^3 x_u^i a_i + a = 0$ $(u = 1,2)$, then x^i also satisfy (10). In fact, we have

$$\sum_{i=1}^{3} x^i a_i + a = (1-h)\left(\sum_{i=1}^{3} x^i a_i + a\right) + h\left(\sum_{i=1}^{3} x_2^i a_i + a\right) = 0.$$

Further, I6 is proven by the fact that, whenever two planes have a common point, the system of the equations of the two planes is compatible, hence it defines a line belonging to the two planes.

Finally, I7 is obtained by showing, using a straightforward computation, that the points (0,0,0), (1,0,0), (0,1,0) and (0,0,1) are not in the same plane.

Hence, the defined space is a model of the affine incidence axioms.

As far as the parallel axiom is concerned, we already remarked that, like in classical analytic geometry, a line is defined by one of its points and the direction coefficients, and that two lines are parallel iff their direction coefficients are proportional. It follows that, given a line and an exterior point, we have just one line defined by that point and by the direction coefficients of the given line, which proves the parallel axiom II.

Hence, we defined an affine space and now we must show that its scalar field is isomorphic to K.

Let us choose the origin at the point $O = (0,0,0)$ and let $P = (p^i)$, $Q = (q^i)$ $(i = 1,2,3)$ be two other points. Together they form an O-pair iff they are collinear with O and $Q \neq O$. Suppose, for instance, that $q^3 \neq 0$, and consider the line

$$x^1 = x^3 a + b, \quad x^2 = x^3 c + d$$

passing through O, P and Q. Then we get $b = d = O$ and the compatibility conditions

$$p^1 = p^3 (q^3)^{-1} q^1, \quad p^2 = p^3 (q^3)^{-1} q^2.$$

Or, if we put $p^3 (q^3)^{-1} = k$, we get

$$p^i = kq^i \qquad (i = 1,2,3). \tag{16}$$

Conversely, the conditions (16) assure the collinearity of O,P,Q. Moreover, remark that the element $k \in K$ is uniquely defined.

We leave to the reader to prove, by elementary analytic geometry, that two O-pairs are similar iff they have the same corresponding scalar k.

Hence, the correspondence

$$(P,Q)_O \leftrightarrow k, \quad k \leftrightarrow ((kq^i),(q^i))$$

defines a bijection between the scalar field of the constructed affine space and the given field K.

This bijection preserves products as we see, using the coresponding definitions, by:

$$((kq^i),(q^i)) \cdot ((k'q^i),(q^i))$$

$$= ((kk'q^i),(k'q^i)) \cdot ((k'q^i),(q^i))$$

$$= ((kk'q^i),(q^i)).$$

For a similar result on sums we have to deal first with vectors.

Using again the above notation, let us consider two vectors $\overrightarrow{PQ}$ and $\overrightarrow{P'Q'}$. By elementary analytic geometry, we easily arrive at the fact that they are equipollent iff

$$q^i - p^i = q'^i - p'^i,$$

which shows that the vectors can be identified with ordered triples of elements of K, namely, the vector defined by two points will be the triple of the differences of their coordinates: $\overrightarrow{PQ} = (q^i - p^i)$.

Now, from the definition of the sum of two vectors, we see that this sum is obtained by adding the components of the vectors.

Next, from the definition of the scalar sum, we get

$$((kq^i),(q^i)) + ((k'q^i),(q^i)) = (((k+k')q^i),(q^i)),$$

which shows that the sums are preserved as well, and this completes the proof of Theorem 6.

The affine space constructed in Theorem 6 is called the *affine space over the field* K.

Let us make also a few other comments. First, the theorem proven obviously shows the consistency of the axiom system of affine geometry and, by passing to the enlarged space, that of projective geometry, because it is providing explicit models.

For instance, one model is the affine space over the field of congruence classes modulo p (where p is a prime integer). This shows that the affine geometry is noncontradictory if the arithmetic of congruences modulo p is such.

Moreover, Theorem 6 defines actually all the affine spaces, up to isomorphism, because, for any affine space, going over to coordinates is just an isomorphism onto the model defined by Theorem 6 over the scalar field of that space.

In other words, if we add to Axioms I and II the requirement that the scalar field be some given field, we get a categorical axiom system which defines, up to isomorphism, a unique affine space.

At this point, we can also prove the independence of the Pappus-Pascal Theorem on the axioms of affine geometry. In fact, this theorem is equivalent with the commutativity of the scalar field and it does not hold in an affine space over a noncommutative field, e.g., the quaternion field.

If, in the proof of Theorem 6, we take the ordered pairs (x^1, x^2) of elements of K as points, and loci defined by equations $x^1 a + x^2 b + c = 0$ as lines (a, b must not vanish simultaneously), we get the *affine plane over the field* K. By remarking that this plane may be identified with the plane $x^3 = 0$ of the affine space over K, it follows that it can be embedded in an affine three-dimensional space, whence, it is an Arguesian plane. If we take again a noncommutative field K, we see that the Pappus-Pascal Theorem is independent of the axioms of plane Arguesian affine geometry.

§5. AFFINE TRANSFORMATIONS

a. Characteristic Properties of Affine Transformations

The last section of this chapter is devoted to a geometrically important matter: point transformations preserving the affine properties in an affine space.

1. DEFINITION. An affine transformation or affine automorphism is a bijective point transformation of the affine space onto itself, which sends lines onto lines and planes onto planes.

The name is clearly justified by the fact that the affine structure of the space is the structure defined by lines and planes. From the definition we see (the details are left for the reader to work out) that the inverse of an affine transformation is also affine and the product of two affine transformations is affine too. Hence we have

2. THEOREM. The set of the affine transformations of the space onto itself is a group, under the usual composition law.

The group of this theorem is denoted by A and is called the *affine group* or the *basic group of the affine space* because the affine properties are those properties of the space which are invariant by the above group of transformations. In fact, the notions of a line and a plane as well as the incidence relations of these elements are invariant, by Definition 1. Next, by reductio ad absurdum, we see that two parallel lines are sent, by an affine transformation, to parallel lines, whence the various relations of lines and planes are preserved. Consequently, affine geometry is just the study of the notions and properties which are preserved by the affine transformation group.

In geometry, this idea is generalized and the various geometric spaces can be defined as sets endowed with a group of one-to-one point transformations, which is called the basic group of the space.

The geometry of the space is then the study of the invariants of the corresponding group. This definition of geometries was given by Felix Klein in his dissertation at the University of Erlangen (1872) and it is known in the literature as the *Erlangen Program*. Hence, affine geometry belongs to the Erlangen Program. The same is true for projective geometry, whose fundamental group is the group of the *projective transformations*, which are defined like the affine transformations (i.e., they have to send lines onto lines and planes onto planes) but in projective space. We shall see later on that Euclidean geometry as well has a basic group which will be suitably defined.

Coming back to the affine transformations, we should note first that, actually, Definition 1 requires too much and that we have

3. THEOREM. If in an affine space the strong incidence axioms hold, then, in order that a one-to-one point transformation of the space onto itself be an affine transformation, it is necessary and sufficient that this transformation send every triple of collinear points to a triple of collinear points. If only the usual incidence axioms are considered, we must add to this condition the requirement that parallelism of lines must be preserved.

The necessity part is obvious, hence let us prove the sufficiency.

First, let A,B,C,D be four coplanar points. If three of them are collinear, the images of the four points, by a transformation T which preserves collinearity, are again coplanar. Now, suppose that every three of the four given points are noncollinear and let A', B',C',D' be their images under T. Consider (Fig. 32) the parallel through D to AC and its intersection point F with AB. Next, using the strong incidence axioms, take on AF a third point, different from A and F, denoted by P, and put $Q = DP \cap AC$. Q exists because $DP \neq DF$. It follows that D' is collinear with $P' = T(P)$ and $Q' = T(Q)$ and, hence, D' belongs to the plane A',B',C'. Thus, in all the cases, four coplanar points are sent by T to four coplanar points.

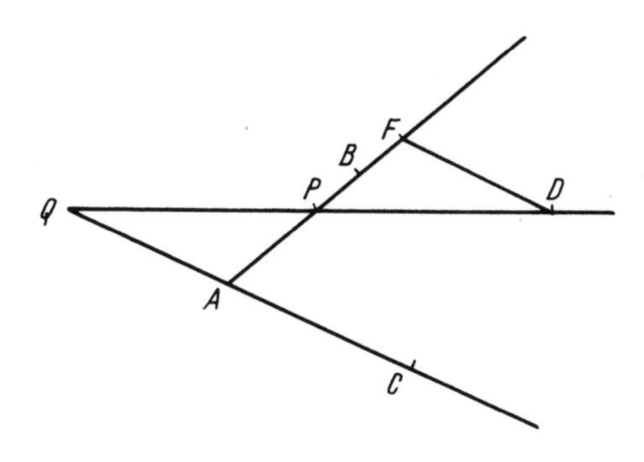

Fig. 32

If we are not allowed to use the strong incidence axioms, the existence of P is not assured, and the conclusion may be reached only through the supplementary condition of the preservation of paralelism of lines, which assures that the image of the line FD is parallel to $A'C'$, whence it is in the plane A',B',C'. The supplementary condition is actually needed and we propose to the reader to provide an example of a one-to-one point-transformation in the affine space over the field of the congruence classes modulo 2, which does not preserve parallelism. (The collinearity will be always preserved, because a line has no more than two points.)

However, if we consider affine automorphisms in an Arguesian plane, this supplementary condition is not needed because the only case where the strong incidence axioms do not hold is that where the scalar field is the field of congruence classes modulo 2 and, in this case the plane has exactly four points any three of which are noncollinear. Thus, every permutation of this plane preserves parallelism.

Hence, if the transformation T preserves collinearity, it also preserves coplanarity and sends a plane into a plane.

In the sequel, we shall show that the same T sends noncoplanar points to noncoplanar points. In fact, if the noncoplanar points A,B,C,D would be sent to the coplanar points $A',B',C',D' \in \alpha$, where α is some plane, then, clearly, all the points of the edges and faces of the tetrahedron $ABCD$ are sent into α. Next, if M is an arbitrary point, join M and D and, if MD intersects the plane ABC in E, the image $T(M)$ is on the image of the line DE by T, hence $T(M) \in \alpha$. If $MD \parallel ABC$ and we have the strong incidence axioms, we replace DE by $D'E'$, where D' is on CD, and we derive the same result. If we are not allowed to use the strong incidence axioms, we get the same conclusion, if we require parallelism of lines to be preserved. Hence, in all the cases, the whole space is sent by T into α which contradicts the fact that T is bijective.

From the above-mentioned property we can now derive that T sends noncollinear points to noncollinear points. Since, if A,B,C are noncollinear and their images A',B',C' are on a line, then, while taking D noncoplanar with A,B,C, we have that A',B',C',D' are in the same plane, in contradiction to what we proved earlier.

It is now trivial to see that T maps bijectively lines onto lines and planes onto planes, which completes the proof of Theorem 3.

b. Special Affine Transformations

Now, we shall consider some particular types of affine transformations which give new interpretations to the fundamental algebraic structure of the space.

4. DEFINITION. An affine transformation is called *special* if it sends every line either onto the same line or onto a parallel line.

Let T be a special affine transformation, A a point of the space, and $A' = T(A)$ its image. If $B \in AA'$, because $AB \underline{\parallel} A'B'$, then the lines AB and $A'B'$ coincide and $B' \in AA'$. From which it follows that T preserves every line which joins a point to its image. (Such a line is fixed, but not every point of it is fixed.)

Another straightforward remark is that, if A is a *fixed point* of the transformation T (i.e., $T(A) = A$), every line through A is preserved by T (because it cannot be sent to a parallel line).

Consequently, if T has at least two fixed points, A, B, T is the identity, because every point $C \bar{\in} AB$ is given as the intersection of the fixed lines AC, BC and must therefore be a fixed point, and a point $C' \in AB$ is also fixed because $C' = AC' \cap CC'$, with C being defined as above.

It follows that there are just the following types of special affine transformations: a) without fixed points, b) with exactly one fixed point, c) the identical transformation.

5. DEFINITION. A special affine transformation which has either no fixed points or all points fixed is called a *translation*.

Let T be a translation without fixed points and A, B two points such that $AB \parallel A'B'$ (as usual, "$'$" denotes the images under T). Then, AA' and BB' are coplanar lines and have no common point since such a point would be fixed. Hence $AA' \parallel BB'$ and we see that $\overrightarrow{AA'} \approx \overrightarrow{BB'}$. Similarly, if the lines AB and $A'B'$ coincide, we may take C not on AB and get $\overrightarrow{AA'} \approx \overrightarrow{CC'}$, $\overrightarrow{CC'} \approx \overrightarrow{BB'}$. Thus, we have always $\overrightarrow{AA'} \sim \overrightarrow{BB'}$ which implies:

6. PROPOSITION. For a translation, the oriented segments joining corresponding points are all equipollent. A translation is uniquely defined by a pair of corresponding points.

In view of this proposition, it is natural to denote a translation by $T_{AA'}$, indicating a pair of corresponding points. In addition, because the segments $\overrightarrow{AA'}$ define a unique vector $\bar{v}$, we can say that the translation is associated with the vector $\bar{v}$ and we can denote it by $T_{\bar{v}}$.

For the identical translation, we take $\bar{v} = \bar{0}$.

Conversely, consider a point transformation in space, denoted by $T_{\bar{v}}$, associated with the vector $\bar{v}$, and sending every point A to the point A' defined by the condition $\overrightarrow{AA'} = \bar{v}$. Clearly, $T_{\bar{v}}$ has the inverse transformation $T_{-\bar{v}}$ (similarly defined), hence $T_{\bar{v}}$ is bijective. We shall see that $T_{\bar{v}}$ is a translation.

In fact, if A,B,C are three collinear points on a line which is parallel to $\bar{v}$, $T_{\bar{v}}$ sends them to points of the same line, i.e., to collinear points. If A,B,C are again collinear but their line is not $\parallel \bar{v}$, their images are on a line which is parallel to the line ABC. Thus $T_{\bar{v}}$ sends collinear points to collinear points, and every line either to the same line or to a parallel line. From this we can also see that $T_{\bar{v}}$ preserves parallelism. Hence, by Theorem 3 and Definition 4, $T_{\bar{v}}$ is a special affine transformation. Moreover, if $\bar{v} \neq \bar{0}$, $T_{\bar{v}}$ has no fixed points and if $\bar{v} = \bar{0}$, all the points are fixed, which proves the assertion of $T_{\bar{v}}$ being a translation.

Consequently, we have a new characterization of the translations, as the transformations $T_{\bar{v}}$, and, of course, these have all the properties of any affine transformation, e.g., preservation of planes, etc.

Proposition 6 and the analysis above show that there is a one-to-one correspondence between the set V of the vectors of the affine space and the set T of the translations. Moreover, from the definition of the sum of vectors it follows obviously that

$$T_{\bar{u}} \circ T_{\bar{v}} = T_{\bar{u}+\bar{v}}$$

and we deduce

7. THEOREM. T is an abelian group, for the usual composition of transformations, and it is isomorphic to the additive group V of vectors.

8. DEFINITION. A special affine transformation with one fixed point or which has only fixed points is called an *homothety* and the fixed point is called its *centre*.

Let S be a homothety of centre O, then consider the points Q and $P = S(Q)$. These points are collinear with O because the line OQ is sent by S to the same line OQ. Further, if we also have another pair Q', $P' = S(Q')$, on a different line through O, we get, from Definition 4, $PP' \parallel QQ'$, whence $(P',Q')_O \approx (P,Q)_O$. If Q' and P' are on

OQ', we may compare them with some P'',Q'' on another line, from which we get $(P',Q') \sim (P,Q)$. Hence, we have

9. PROPOSITION. If S is a homothety, all the O-pairs $(S(Q),Q)$ are similar. A homothety is uniquely defined by a pair of corresponding points.

Consequently, we can speak of the homothety S_k^O associated with the scalar $k = (S_k^O(Q),Q)_O$.

Conversely, denote by $S_k^O (k \neq 0)$ the point transformation sending every point $Q \neq O$ to the point P defined by $(P,Q) = k$, and which fixes the point O. It is simple to see that this transformation is bijective and sends every line to either a coincident or a parallel line. Implicitly, it preserves the collinearity and the parallelism. Hence, this is a special affine transformation with just one fixed point if $k \neq 1$ and it is identical if $k = 1$. Therefore, S_k^O is always a homothety.

Thus, we have a new characterization of homotheties as the transformations S_k^O, which implies that they have the usual affine properties of preserving planes, etc.

Let us also add the *degenerate* transformation S_0^O which sends all the points to O. Then, we denote $O_O = \{S_k^O\}$ and we shall prove

10. THEOREM. The set O_O of all the homotheties of centre O may be provided with a field structure which makes it isomorphic to the scalar field K of the corresponding affine space.

In fact, it follows from the discussion that we have a natural bijection $\varphi : K \to O_O$, whence, by defining the sum in O_O such that it will be compatible with φ, we shall have an isomorphism between the additive structures of K and O_O. Next, the product by zero $= S_0^O$ in O_O is defined to give always the result 0, and the product of two arbitrary $(\neq 0)$ homotheties in O_O is defined by the usual composition of maps. We clearly have

$$(S^O_{k_1} S^O_{k_2}(Q), S^O_{k_2}(Q))(S^O_{k_2}(Q), Q) = (S^O_{k_1} S^O_{k_2}(Q), Q) ,$$

which shows that φ preserves the products and it is an isomorphism of the corresponding multiplicative structures as well. Implicitly, $\mathcal{O}_O$ is a field and with it Theorem 10 is proved.

11. COROLLARY. The multiplicative group of the homotheties of centre O is commutative iff the space is a Pappian space (i.e., the Pappus-Pascal Theorem holds).

c. The Structure of the Affine Group

We shall proceed now with some properties regarding the structure of the affine transformation group.

12. THEOREM. The group T of the translations is a normal subgroup of the affine group A. Also, if A_{sp} is the set of the special affine automorphisms, A_{sp} is a subgroup of A and T is a normal subgroup of A_{sp}.

Indeed, if T is a nonidentical translation and A an affine transformation, ATA^{-1} is a special affine automorphism with no fixed points because, if $ATA^{-1}(M) = M, A^{-1}(M)$ would be a fixed point of T which is impossible. Hence ATA^{-1} is also a translation and the first part of Theorem 12 is proven. The second part is trivial.

Let us also note that in relation to Theorem 12, the operation of multiplying a vector by a scalar admits an interpretation by using translations and homotheties. Namely, $k\bar{v}$ is the vector corresponding to the translation $S_k T_{\bar{v}} S_{k^{-1}}$. This follows in a straightforward manner from the corresponding definitions.

Also, Theorem 12 suggests considering the quotient group A/T and it is interesting that this group is isomorphic to a subgroup of the affine group.

In fact, denote by A_O the subset of affine transformations which have the origin O as a fixed point and which obviously form a subgroup of A.

If $A_i \in \mathsf{A}_o$ $(i = 1,2)$, a relation $A_1 = A_2 T$ is impossible for $T \in \mathsf{T}$ ($T \neq \text{Id.}$) because then $T = A_2^{-1} A_1$ has the fixed point O. On the other hand, we can immediately find in every class of A/T one and only one element of A_o, namely, if A is any element of that class and $O' = A^{-1}(O)$, the required element is $AT_{OO'}$. By associating then to every class its element in A_o, we get an epimorphism $\mathsf{A}/\mathsf{T} \to \mathsf{A}_o$. If the class of A is sent to the identity, we have $AT_{OO'} = \text{id.}$, which implies that A is a translation. Hence, the epimorphism is also a monomorphism and it is just the required isomorphism.

The group A_o defined above is called the *centroaffine group*, and it defines an interesting geometry belonging to the Erlangen Program.

Now, we shall prove

13. THEOREM. Any affine transformation has a unique decomposition as a product of a translation and a centroaffine transformation.

Indeed, using the notation above, if A is an affine transformation, then $A' = AT_{OO'}$ is centroaffine. Hence, we have $A = A'T_{O'O}$ and this decomposition is unique since two centroaffine transformations cannot be equivalent with respect to T.

Another interesting fact comes from the following considerations. Since an affine transformation preserves parallelism, it follows immediately that it also preserves the equipollence of oriented segments. Hence, such a transformation induces a transformation on the vector space V and the affine group acts also as a transformation group on V.

But this action is not *effective*, which means that there are non-identical affine transformations which leave all the vectors unchanged. Indeed, if $T_{\bar{v}}$ is the translation associated with the vector $\bar{v}$, and $T_{\bar{v}}(\overrightarrow{AB}) = \overrightarrow{A'B'}$, it follows $\overrightarrow{AA'} = \overrightarrow{BB'} = \bar{v}$ and, by the parallelogram property (Corollary 3.9), we get $\overrightarrow{A'B'} = \overrightarrow{AB}$. Thus, $T_{\bar{v}}$ leaves all the vectors unchanged. Conversely, if T is a transformation which has this property, we have $T(\overrightarrow{AB}) = \overrightarrow{A'B'} \sim \overrightarrow{AB}$, whence, using again the mentioned parallelogram property, $\overrightarrow{AA'} = \overrightarrow{BB'}$, and T is just the translation $T_{AA'}$. Hence, we proved

14. PROPOSITION. A bijective transformation of the space is a translation iff it leaves all the vectors unchanged.

From this proposition and Theorem 13 we get

15. THEOREM. There is an effective action of the group A/T, i.e., of the centroaffine group A_o, on the space V. The respective transformations are automorphisms of the additive group structure of V.

In fact, we saw that A acts on V, and T is the subgroup leaving V invariant. It follows that all the elements of a class of A/T act in the same way on V. Hence, we get an action of A/T as a transformation group on V and no non-identical transformation leaves all the vectors unchanged (which is just the effectiveness property). This proves the first part of the theorem. If $T \in A_o$, $T(\bar{v})$ will be obtained by taking $\bar{v} = \overrightarrow{OA}$ and next $T(\bar{v}) = \overrightarrow{OT(A)}$.

The last part of Theorem 15 may be obtained as follows. Let T be an affine transformation. Then

$$T(\overrightarrow{AB} + \overrightarrow{BC}) = T(\overrightarrow{AC}) = \overrightarrow{A'C'} = \overrightarrow{A'B'} + \overrightarrow{B'C'},$$

where $T(A) = A'$, $T(B) = B'$, $T(C) = C'$. Thus, T is an endomorphism of the additive group V and, because it has the inverse T^{-1}, it is an automorphism.

Let us deal now with the behavior of the above transformations when a vector is multiplied by a scalar.

Let $\bar{v} \neq \bar{0}$ be a fixed vector, and A a centroaffine transformation with respect to a fixed origin O. Because A preserves point collinearity, there will be for every scalar k a uniquely associated scalar m such that $A(k\bar{v}) = mA(\bar{v})$. Hence, we get a map $m : K \to K$ (where, as usually, K denotes the scalar field).

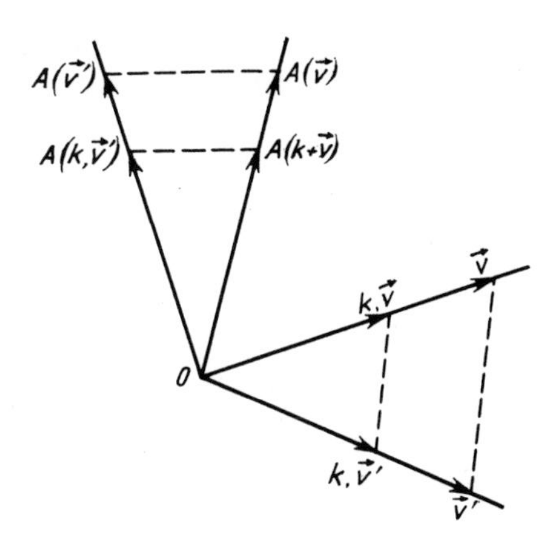

Fig. 33

This map is easily seen to depend on the transformation A only, and not on the used vector $\bar{v}$. In fact, if we take another vector $\bar{v}'$, since A preserves parallelism and in view of the definition of the multiplication of vectors with scalars, we get (Fig. 33) $A(k\bar{v}') = mA(\bar{v}')$ with the same m as before.

Now, the following computations hold

$$m(k_1 + k_2)A(\bar{v})$$
$$= A[(k_1 + k_2)\bar{v}] = A(k_1\bar{v} + k_2\bar{v})$$
$$= A(k_1\bar{v}) + A(k_2\bar{v}) = m(k_1)A(\bar{v}) + m(k_2)A(\bar{v}), \quad m(k_1k_2)A(\bar{v})$$
$$= A(k_1(k_2\bar{v})) = m(k_1)A(k_2\bar{v})$$
$$= m(k_1)m(k_2)A(\bar{v}),$$

and we obtain

$$m(k_1 + k_2) = m(k_1) + m(k_2),$$

$$m(k_1k_2) = m(k_1)m(k_2),$$

which shows that m is a homomorphism.

Using again the preservation of the collinearity, it also follows that, for a given scalar m, there is some scalar k such that $A^{-1}(mA(\bar{v})) = k\bar{v}$, i.e., such that $m(k) = m$.

Hence, $m : K \to K$ is an epimorphism of fields, i.e., it is an automorphism (see §2 of the Introduction).

It follows

16. THEOREM. Every centroaffine transformation and, hence, every affine transformation (see Theorem 13) has a well defined associated automorphism of the scalar field such that, in the above notation, $A(k\bar{v}) = m(k)A(\bar{v})$.

This theorem implies a classification of the affine transformations by the corresponding automorphisms of the scalar field such that

$$A = \bigcup_{m} A^{(m)},$$

where m runs over the set of the automorphisms of K.

As examples, we shall consider the case of the real field R and of the complex field C.

If $\varphi : R \to R$ is an automorphism, we have $\varphi(1) = 1$. Next $\varphi(n) = \varphi(1 + \ldots + 1) = n$ and, since

$$1 = \varphi(1) = \varphi\left(n \cdot \frac{1}{n}\right) = \varphi(n)\,\varphi\left(\frac{1}{n}\right),$$

$\varphi(1/n) = 1/n$. Then, using again the compatibility with the sum, we get $\varphi(m/n) = m/n$ for every rational nonnegative number m/n. Now, from $\varphi(-n/m) + \varphi(n/m) = 0$, we get $\varphi(-n/m) = -n/m$, hence the restriction of φ to the rational numbers is the identical transformation.

Further, let $\alpha > 0$ be a real number. Then we can write $\alpha = a^2$ and $\varphi(\alpha) = (\varphi(a))^2$, which shows that $\varphi(\alpha) > 0$. Analogously, $\alpha < 0$ implies $\varphi(\alpha) < 0$, whence φ is a monotonically increasing function with respect to the usual order of the real numbers.

An irrational number x may be defined as the common point of a sequence of rational intervals strictly included in one another and whose lengths tend to 0, i.e.,

$$r' < x < r''.$$

By applying φ, we get in view of the above facts

$$r' < \varphi(x) < r'',$$

which shows that $\varphi(x)$ is common to just the same sequence of intervals like x. Hence $\varphi(x) = x$ and we see that *the identity is the only automorphism of the real field R.*

This means that in the affine space over the real field R we have but one class of affine transformations and these act as linear transformations on the corresponding space of vectors. In this case, the centroaffine group identifies with the so-called *real linear group*.

As for the complex field C, it is interesting to obtain those automorphisms $\varphi : C \to C$, such that the φ/R sends R to R and, therefore, it is the identity map. Then, we must have $\varphi(a + ib) = \varphi(a) + \varphi(i)\varphi(b) = a + \varphi(i)b$ and φ is defined if we know $\varphi(i)$. Since $\varphi(i^2) = \varphi^2(i) = -1$, we must have $\varphi(i) = \pm i$ and we see that there are only two types of required automorphisms: the identity map and the complex conjugation map. (We must note that the condition φ/R sends R to R is essential here.) The two corresponding types of affine transformations are of importance in complex affine geometry.

d. Determination of Affine Transformations

Now, as an application of Theorem 16, we shall prove the following result called *the main theorem* for affine transformations.

17. THEOREM. Let A_i and A_i' $(i = 1,2,3,4)$ be two systems of non-coplanar points. Then, in every class $A^{(m)}$ of affine

transformations there is a unique transformation T such that $A_i' = T(A_i)$.

Indeed, if T exists and M is an arbitrary point such that

$$\overrightarrow{A_1M} = a\overrightarrow{A_1A_2} + b\overrightarrow{A_1A_3} + c\overrightarrow{A_1A_4}$$

and $M' = T(M)$, we have

$$\overrightarrow{A_1'M'} = m(a)\overrightarrow{A_1'A_2'} + m(b)\overrightarrow{A_1'A_3'} + m(c)\overrightarrow{A_1'A_4'}.$$

Hence M' is well defined and T is unique.

Now, to establish the existence of T it suffices to show that the transformation $M' = T(M)$ defined by the two formulas above is affine.

This will be done by using Theorem 3. Let M_h $(h = 1,2,3)$ be three collinear points, i.e., $\overrightarrow{M_3M_1} = k\overrightarrow{M_2M_1}$. Then M_h' is defined by putting

$$\overrightarrow{A_1M_h} = a_h\overrightarrow{A_1A_2} + b_h\overrightarrow{A_1A_3} + c_h\overrightarrow{A_1A_4}$$

and

$$\overrightarrow{A_1'M_h'} = m(a_h)\overrightarrow{A_1'A_2'} + m(b_h)\overrightarrow{A_1'A_3'} + m(c_h)\overrightarrow{A_1'A_4'}.$$

This yields

$$\overrightarrow{M_2M_1} = (a_1 - a_2)\overrightarrow{A_1A_2} + (b_1 - b_2)\overrightarrow{A_1A_3} + (c_1 - c_2)\overrightarrow{A_1A_4},$$

$$\overrightarrow{M_3M_1} = (a_1 - a_3)\overrightarrow{A_1A_2} + (b_1 - b_3)\overrightarrow{A_1A_3} + (c_1 - c_3)\overrightarrow{A_1A_4},$$

and, since m is an automorphism,

$$\overrightarrow{M_2'M_1'} = m(a_1 - a_2)\overrightarrow{A_1'A_2'} + m(b_1 - b_2)\overrightarrow{A_1'A_3'} + m(c_1 - c_2)\overrightarrow{A_1'A_4'},$$

$$\overrightarrow{M_3'M_1'} = m(a_1 - a_3)\overrightarrow{A_1'A_2'} + m(b_1 - b_3)\overrightarrow{A_1'A_3'} + m(c_1 - c_3)\overrightarrow{A_1'A_4'}.$$

The collinearity of the given points is equivalent to

$$a_1 - a_3 = k(a_1 - a_2), \quad b_1 - b_3 = k(b_1 - b_2), \quad c_1 - c_3 = k(c_1 - c_2),$$

which implies

$$m(a_1 - a_3) = m(k)m(a_1 - a_2)$$

and two similar relations for b and c. But these means $\overrightarrow{M_3'M_1'} = m(k)\overrightarrow{M_2'M_1'}$, i.e., M_h' are collinear points.

Hence, T preserves collinearity. Moreover, the above proof shows that, actually, T preserves the collinearity of the vectors, which means that T also preserves the line parallelism. Hence, T is an affine transformation, which completes the proof of Theorem 17.

Theorem 17 also admits the following reformulation:

18. THEOREM. Let $\{O, \bar{e}_i\}$ and $\{O', \bar{e}_i'\}$ $(i = 1,2,3)$ be two arbitrary affine frames. Then, there is a unique affine transformation T in every class $A^{(m)}$ such that $O' = T(O)$, $\bar{e}_i' = T(\bar{e}_i)$.

In fact, we have only to apply Theorem 17 to the systems of points defined by the origins and the ends of the basic vectors constructed, starting from the corresponding origins. It is also obvious that Theorem 17 follows from Theorem 18 if we use the frames defined by the two given systems of points.

It is interesting to show that the equations of the transformation T of Theorem 18 have very simple expressions if we use coordinates with respect to the two corresponding frames. In fact, if M has the coordinates x^i with respect to the first frame, we have $\overrightarrow{OM} = \sum_{i=1}^{3} x^i \bar{e}_i$, whence

$$\overrightarrow{O'M'} = \sum_{i=1}^{3} m(x^i)\bar{e}_i' = \sum_{i=1}^{3} x'^i \bar{e}_i',$$

and the required equations are

$$x'^i = m(x^i) \qquad (i = 1,2,3).$$

Hence, any affine transformation has the simple equations stated above, when expressed in corresponding frames.

For the real field R, $m = \text{id.}$ and the transformation T of Theorems 17 and 18 is unique. With respect to corresponding frames, the equations of the transformation are $x'^i = x^i$.

Let us also derive the equations of an affine transformation with respect to a fixed frame $\{O, \bar{e}_i\}$.

If we denote the transformation by T and put $O' = T(O)$, and if p^i are the coordinates of O' with respect to that frame and

$$T(\bar{e}_i) = \bar{e}'_i = \sum_{j=1}^{3} p^j_i \bar{e}_j,$$

then, for $M(x^i)$ and $T(M) = M'(x'^i)$, we shall have

$$T(\overrightarrow{OM}) = \overrightarrow{O'M'} = \overrightarrow{OM'} - \overrightarrow{OO'} = \sum_{i=1}^{3} (x'^i - p^i)\bar{e}_i$$

$$= T\left(\sum_{i=1}^{3} x^i \bar{e}_i\right) = \sum_{j,i=1}^{3} m(x^i) p^j_i \bar{e}_j.$$

It follows that

$$x'^i = \sum_{j=1}^{3} m(x^j) p^i_j + p^i, \tag{1}$$

which are just the required equations.

If we compare equations (1) and equations 4.(7), we see that the elements of $A^{(\mathrm{id})}$ have the same analytic expression as the frame transformations.

Another interesting property of $A^{(\mathrm{id})}$ is that this is a subgroup of A, which is, generally, not true for the other $A^{(m)}$. Hence, $A^{(\mathrm{id})}$ defines a geometry belonging to the Erlangen Program and, in this geometry the scalars are invariant by the transformations of the basic group. Usually, the scalar $(P,Q)_O$ is called the *ratio* of the triple of collinear points O,P,Q and is denoted by $(P,Q;O)$ or by PO/QO. Over the real field, this is true for the whole affine group.

If T is a translation, we have $\overrightarrow{O'M'} \sim \overrightarrow{OM}$ and the calculation done for (1) gives now

$$x'^i = x^i + p^i \tag{2}$$

or, in vector form,

$$\bar{r}' = \bar{r} + \bar{p}, \tag{3}$$

which shows that $T \subseteq A^{(\mathrm{id})}$.

If T is a homothety with centre O, then $O' = T(O) = O$ and $\overrightarrow{OM'} = k\overrightarrow{OM}$, whence the equations of the homothety will be

$$\bar{r}' = k\bar{r} \tag{4}$$

or

$$x'^i = kx^i = (kx^i k^{-1})k. \tag{5}$$

We see that, in this case, the automorphism m is $m(x) = kxk^{-1}$.

If in (1) $p^i = 0$, we get a centroaffine transformation and conversely. Hence, the formulas (1) give the decomposition of the affine transformation as a product of a translation and a centroaffine transformation, as indicated by Theorem 13.

Finally, if in the equations (1), we go over to homogeneous coordinates, we get the equations of the projective transformations and these are of the following form

$$\rho y'^i = \sum_{j=1}^{3} m(y^i)a_j^i,$$

where ρ is an arbitrary non-zero proportionality factor, which appears because of the homogeneous character of the coordinates, and (a_j^i) is a nondegenerate (i.e., invertible) matrix.

We end this section with the remark that, by similar methods, we can study affine transformations between different spaces as well as between lines or planes.

PROBLEMS

1. Define a model of the plane incidence axioms I1, I2, I3 having exactly three points.

2. Define a model of the space incidence axioms I1-I7 having exactly four points. How many lines and how many planes are in this model?

3. Give a detailed proof of the fact that any plane has at least two concurrent lines.

4. How many pairs of skew lines and how many pairs of parallel lines are in the four-point model of Problem 2?

5. Define a model of the plane strong incidence axioms I1, I2', I3, having exactly nine points. Are there also models with fewer points?

6. Define a model of the spatial strong incidence axioms I1, I2', I3-I7, having exactly 27 points. Are there also models with fewer points?

7. Define models of affine planes and spaces having a minimum number of points. Consider both cases, when the usual and the strong incidence axioms are imposed.

8. An affine plane is called *finite* if it has a finite number of points. Prove that in a finite affine plane the number of lines passing through a given point is constant. If this number is $n+1$, we say that the plane is of *order* n.

9. Prove the following properties of the order n of a finite affine plane:

   a) $n \geq 2$;

   b) every line contains n points;

   c) the plane has n^2 points and $n^2 + n$ lines;

   d) there are exactly $n+1$ pencils of parallel lines and each such pencil has n lines.

10. Consider a finite affine plane of order n with the points A_1, $\ldots, A_{n^2}$ and the lines $a_1, \ldots, a_{n^2+n}$. Construct the matrix M of order (n^2, n^2+n) whose entry a_{ij} is 1 if $A_i \in a_j$ and 0 if $A_i \bar{\in} a_j$. M is called the *incidence matrix* of the plane. Prove that $MM' = nI_{n^2} + J_{n^2}$, where M' is the transposed matrix of M, I_{n^2} is the unit matrix of order n^2, and J_{n^2} is the matrix of order n^2 with all entries 1.

11. Give detailed proofs of all the properties of Theorem 2.10.

12. Formulate an independent axiomatic definition of the projective plane and of projective space without using the enlarged affine space. Show that the homogeneous coordinates over fields define analytic models of the projective plane and space.

13. Show that, by choosing an arbitrary plane of the projective space as the locus of the improper elements, one obtains in the space a structure of an enlarged affine space.

14. Let M be a set of elements called *points* in which there is a distinguished family of subsets called *lines*. M is said to be a *configuration* iff two distinct points belong to at most one line. Show that the 10 points and 10 lines encountered in the Desargues Theorem for projective space form a configuration.

15. Let M be a configuration with at least four points such that three of them are never collinear. Define M_1 as the configuration obtained by adding to M, as new lines, all the pairs of non-collinear points of M. Then, define M_2 by adding to M_1, as new points, all the pairs of non-concurrent lines of M_1. Continuing, we add to M_2 new lines, to M_3 new points, etc. Now, take $\Pi = \bigcup_{n=0}^{\infty} M_n$ $(M_0 = M)$, call points of Π all the points of all the M_n and lines of Π all the subsets $L \subset \Pi$ such that $L \cap M_n$ is a line in M_n for all sufficiently large n. Prove that Π is a projective plane. (This will be called the *free projective plane generated by M*.)

16. A configuration M is said to be a *bounded configuration* iff every line has at least three different points and every point

belongs to, at least three lines. Prove that the Desargues configuration of Problem 14 is bounded. Prove that every bounded configuration of the free plane generated by M is actually contained in M.

17. Let M_0 be the configuration consisting of four points and no line. Prove that the free projective plane generated by M_0 is a non-Arguesian plane.

18. Give a detailed proof of the non-Arguesian character of Hilbert's model described in the text.

19. Find an analytical proof for Theorem 2.13 and state from it the Arguesian character of the affine and projective planes over fields.

20. Prove (using Desargues' Theorem) that if the middles of the sides of a triangle exist then the medians of the triangle are either concurrent or parallel and every line which joins two midpoints is parallel to the third side of the triangle.

21. Prove the distributivity relation

$$k(k_1 + k_2) = kk_1 + kk_2$$

for scalars, in the case when one of the following relations holds: $k_1 = 0$, $k_2 = 0$, $k_1 + k_2 = 0$.

22. Give a detailed proof of the fact that the correspondence defined in Section 3 between the scalar fields with different origins is an isomorphism.

23. Find an analytic proof of Theorem 3.20 (Pappus-Pascal).

24. Prove in an independent manner that the set of the vectors of an affine Arguesian plane is a two-dimensional vector space over the scalar field.

25. Show that any two planes of an affine space have the same cardinal number.

26. Let $A_0, A_1, \ldots, A_h$ be $h+1$ points of an affine space. These points are called *independent points* iff A_0A_i $(i = 1,\ldots,h)$ are

independent vectors. Show that, if this happens and if we perform a permuation of the given points, we get again independent points. Show that the maximum number of independent points on a line is 2, in a plane 3, and in space 4.

27. If the points A_σ $(\sigma = 0,\dots,h)$ have the affine coordinates (x^i_σ) $(i = 1,2,3)$, then show that these points are independent iff

$$\operatorname{rank}\begin{pmatrix} x^1_0 & x^2_0 & x^3_0 & 1 \\ \dots & \dots & \dots & \dots \\ x^1_h & x^2_h & x^3_h & 1 \end{pmatrix} = h + 1.$$

28. The quaternion field K is the set of expressions $a + bi + ej + dk$, where addition and multiplication are like for polynomials and

$$ij = -ji = k, \quad jk = -kj = i, \quad ki = -ik = j$$

$$i^2 = j^2 = k^2 = -1.$$

Consider the affine space over K and write down the equations of a line containing two given points and the equation of a plane containing three given points.

29. Consider an affine Arguesian plane and let A,B,C be points on a given line d. Take an arbitrary line $\ell \neq d$ containing C and an arbitrary point $P \in \ell$. Consider the lines $\ell_1 \parallel \ell \parallel \ell_2$ such that $A \in \ell_1$, $B \in \ell_2$, and the points $S = \ell_1 \cap BP$, $R = \ell_2 \cap AP$, $D = d \cap RS$. Show, by an analytic computation that if D exists it depends on the points A,B,C only. The obtained point D is the *harmonic pair* of C with respect to A,B.

30. Show that the scalar field K is of characteristic 2 iff in the construction of Problem 29, $D = C$.

31. Let A,B,C,D be four coplanar points such that any three of them are noncollinear. Suppose that the following intersection points exist: $M = AB \cap CD$, $N = AD \cap BC$, $P = AC \cap BD$, $S = PN \cap AB$. Show that (S,M) and (A,B) are harmonic pairs. Show that the points M,N,P are noncollinear iff the scalar field K is of characteristic $\neq 2$.

32. If AP_2 is the affine plane over the field of integers modulo 2, show that any permutation of AP_2 is an affine transformation. Show that in the space AS_2 over the same field there are permutations which are not affine transformations.

33. Use Theorem 5.17 to show that in the affine plane over the field of integers modulo 3 there are 432 affine automorphisms and in the affine space over the same field there are 303264 affine automorphisms.

34. A bijective correspondence between two planes of an affine space is called an affine transformation between the two planes if it sends lines to lines. Prove that if α and β are two parallel planes and M a point which does not belong to any one of them, then the projection of α onto β from M is an affine transformation. Write down the equations of this transformation with respect to convenient coordinates.

35. Find the fixed points and the fixed lines of an affine transformation of a plane onto itself.

36. A bijective transformation φ such that $\varphi^2 = \mathrm{id}$, $\varphi \neq \mathrm{id}$, is called an *involution*. Find examples of affine involutions.

CHAPTER 2

ORDERED SPACES

At the beginning of the previous chapter, we stated that our aim is to construct a mathematical model of physical space. But, for obvious reasons, this model must have more properties than a general affine space. Indeed, from the intuitive geometry of physical space, we know, among other things, that the points of a line must have an order relation, and from the axioms of affine geometry we cannot get such a relation. Hence, we need some new axioms, a part of which will be studied in this chapter. At this point we make a convention that, throughout this chapter, *the considered spaces are satisfying the strong incidence axioms* (1,§1).

§1. THE ORDER AXIOMS AND THEIR FIRST CONSEQUENCES

a. Linear Order Properties

As it has been mentioned above, we shall add a third group of axioms called *order axioms*. They are characterizing a new *fundamental relation*, which expresses the fact that *a point lies between two other points*. (The relation is fundamental in the sense that it is undefined but characterized by axioms.) If the relation holds for (A, B, C) in the sense that B has to be considered lying between A and C, we shall write A-B-C and we shall also say that we have *the order A-B-C*. The contrary situation will be denoted by $\overline{A\text{-}B\text{-}C}$.

The formulation of the order axiom is

III 1. If A-B-C, then A, B and C are distinct collinear points and we also have C-B-A.

III 2. If A,B,C are three collinear points then one and only one of the relations A-B-C, B-C-A, or C-A-B holds.

III 3. (The Pasch Axiom). Let A,B,C be three noncollinear points, and d be a line in the plane of these points but containing none of them. Then, if d has a point which is between B and C, d also contains, at least, a point between A and B or a point between A and C.

Let us also mention another formulation of the Pasch Axiom. Namely, we already defined a segment as a pair of distinct points and a triangle as a triple of noncollinear points. Now, the points which lie between the ends of a segment will be called *interior points*. The Pasch Axiom may be formulated as follows: *if a line is in the plane of a triangle and contains none of its vertices, and if this line intersects the interior of one of the sides of the triangle, the line also intersects the interior of at least one other side.*

Note that the previous order axioms are referring to plane configurations at most, which means that we have the same order axioms for plane and spatial (three-dimensional) geometry.

1. DEFINITION. The spaces satisfying the strong incidence and order axioms (i.e., the groups I and III) will be called *geometrically ordered spaces*. If the parallel axiom II is also satisfied we have *ordered affine spaces*.

We shall first study geometrically ordered spaces, i.e., without the parallel axiom, and we begin with some simple properties.

2. PROPOSITION. No line intersects the interiors of all the sides of a triangle.

If the contrary were true, we would have some triangle ABC and some line d intersecting its sides in the interior points P,Q,R (Fig. 34). In view of Axiom III 2, we must then have, for instance P-Q-R, whence, for the triangle APR and the line BC, we had a situation contradicting Axiom III 3. The announced proposition follows thereby.

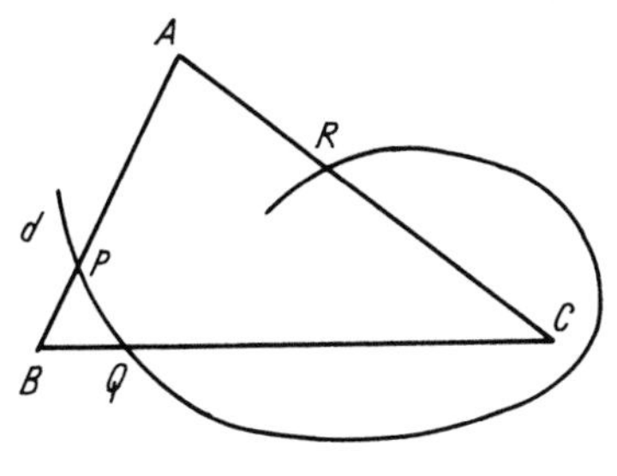

Fig. 34

3. COROLLARY. Given a triangle and a line in its plane which contains none of its vertices, then either the line intersects two sides of the triangle in interior points, or it intersects none.

4. PROPOSITION. Let A,B,C,D be four collinear points and suppose that we have either $\overline{A\text{-}D\text{-}C}$ and $\overline{B\text{-}D\text{-}C}$ or A-D-C and B-D-C. Then we have $\overline{A\text{-}D\text{-}B}$.

Indeed, consider, for instance, the first mentioned situation (Fig. 35). Let E be a point not on AB and consider a third point on CE. By Axiom III 2, it has one of the positions H_1, H_2 or H_3 of Fig. 35. If H_1 exists, take the triangle AH_1C and the line DE and use Axiom III 3. We get some point F such that A-F-H_1. Similarly, with the line DE and the triangle BH_1C we get a point G such that B-G-H_1. Finally, with DE and the triangle ABH_1 we have, by Proposi-

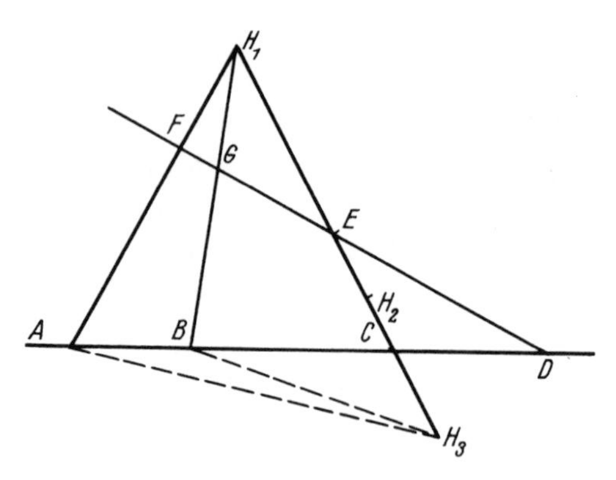

Fig. 35

tion 2, $\overline{A\text{-}D\text{-}B}$, which had to be proved. Now, if H_3 exists, using the line DE and similar triangles like for the previous case, we get in view of Corollary 3 that DE meets none of their sides in interior points and we obtain again the desired conclusion. Finally, if the point H_2 of CE is the one whose existence is known, we change the role of the points E and H in the proof above, and also obtain $\overline{A\text{-}D\text{-}B}$. Hence, the first part of Proposition 4 is proved. The proof of the second part is similar and we leave it to the reader to carry out.

The method of the above proof, which consists in choosing an exterior point and a repeated use of the Pasch Axiom, may be used to obtain many other relations between different points of a line. However, we prefer to state directly the most general order relations.

5. THEOREM. The set of the oriented segments of a line may be uniquely decomposed into two disjoint classes which are non void and such that: a) two segments of the form $\overrightarrow{AP}$, $\overrightarrow{AQ}$ (with the same origin) are in different classes iff $P\text{-}A\text{-}Q$ and in the same class if $\overline{P\text{-}A\text{-}Q}$; b) for every (A,B), the segments $\overrightarrow{AB}, \overrightarrow{BA}$ are in different classes.

For this theorem we shall give the proof of [15].

Suppose that a decomposition satisfying a), b), exists. Then, the classes are non-void (because on the line we have at least one oriented segment and its opposite) and disjoint because of condition a).

Now, let $\overrightarrow{AP}$ and $\overrightarrow{BQ}$ be two arbitrary segments with distinct origins $A \neq B$ and consider the two auxiliary pairs $(\overrightarrow{AP},\overrightarrow{AB})$, $(\overrightarrow{BQ},\overrightarrow{BA})$. In view of condition b) and using the reductio ad absurdum, we see that $\overrightarrow{AP}$ and $\overrightarrow{BQ}$ belong to different classes iff either both auxiliary pairs contain segments of the same class or both auxiliary pairs contain segments of different classes.

It follows that our decomposition is unique because, if one had more than one decomposition with the properties a), b), and, if in one of them $\overrightarrow{AP},\overrightarrow{BQ}$ are in different classes and in another they are in the same class, a similar situation would happen for at least one of the auxiliary pairs, e.g., $(\overrightarrow{AP},\overrightarrow{AB})$. But then, condition a) implies first $P\text{-}A\text{-}B$ and then $\overline{P\text{-}A\text{-}B}$, which is contradictory.

Further, the above characterization of pairs belonging to different classes by means of auxiliary pairs provides also the way of proving the existence of the announced decomposition.

Actually, condition a) and the mentioned characterization suggest associating to each pair of oriented segments a symbol taking the values ±1 according to the following definition

$$\delta(\overrightarrow{AP},\overrightarrow{AQ}) = \begin{cases} -1 & \text{if} \quad P\text{-}A\text{-}Q, \\ +1 & \text{if} \quad \overline{P\text{-}A\text{-}Q}, \end{cases} \tag{1}$$

$$\delta(\overrightarrow{AP},\overrightarrow{BQ}) = -\delta(\overrightarrow{AP},\overrightarrow{AB})\,\delta(\overrightarrow{BQ},\overrightarrow{BA}). \tag{2}$$

From this definition, we also get

$$\delta(\overrightarrow{AB},\overrightarrow{AB}) = -\delta(\overrightarrow{AB},\overrightarrow{BA}) = 1, \tag{3}$$

$$\delta(\overrightarrow{AP},\overrightarrow{BQ}) = \delta(\overrightarrow{BQ},\overrightarrow{AP}) = \frac{1}{\delta(\overrightarrow{AP},\overrightarrow{BQ})}. \tag{4}$$

Next, we fix some segment $\overrightarrow{CR}$ and put in class I all those segments $\overrightarrow{AP}$ for which $\delta(\overrightarrow{AP},\overrightarrow{CR}) = 1$ and in class II all the segments $\overrightarrow{BQ}$ such that $\delta(\overrightarrow{BQ},\overrightarrow{CR}) = -1$. We have to show that the classes I, II satisfy properties a) and b) of the theorem.

To arrive at this result, we shall first derive the condition which characterizes pairs of segments of different classes and, to do this, we shall calculate the expression

$$E = \delta(\overrightarrow{AP},\overrightarrow{BQ})\,\delta(\overrightarrow{BQ},\overrightarrow{CR})\,\delta(\overrightarrow{CR},\overrightarrow{AP}). \tag{5}$$

where $\overrightarrow{CR}$ is the fixed segment and $\overrightarrow{AP},\overrightarrow{BQ}$ are arbitrary oriented segments.

We have three possible situations: 1° $A = B = C$, 2° $A = B \neq C$, 3° $A \neq B \neq C \neq A$.

In case 1°, Proposition 4 shows that either all of the factors of E are +1 or two of them are -1 and the third is +1. Hence, always, $E = 1$.

In case 2° we have

$$E = \delta(\overrightarrow{AP},\overrightarrow{AQ})\,\delta(\overrightarrow{AQ},\overrightarrow{AC})\,\delta(\overrightarrow{CR},\overrightarrow{CA})\,\delta(\overrightarrow{CR},\overrightarrow{CA})\,\delta(\overrightarrow{AP},\overrightarrow{AC})$$

$$= \delta(\overrightarrow{AP},\overrightarrow{AQ})\,\delta(\overrightarrow{AQ},\overrightarrow{AC})\,\delta(\overrightarrow{AC},\overrightarrow{AP}) = 1,$$

where the last equality comes from case 1°.

Finally, in case 3°, we have

$$E = -\delta(\overrightarrow{AP},\overrightarrow{AB})\,\delta(\overrightarrow{BQ},\overrightarrow{BA})\,\delta(\overrightarrow{BQ},\overrightarrow{BC})\,\delta(\overrightarrow{CR},\overrightarrow{CB}) \cdot \delta(\overrightarrow{CR},\overrightarrow{CA})\,\delta(\overrightarrow{AP},\overrightarrow{AC}).$$

But, from case 1°, we easily obtain $\delta(\overrightarrow{AP},\overrightarrow{AB})\,\delta(\overrightarrow{AP},\overrightarrow{AC}) = \delta(\overrightarrow{AB},\overrightarrow{AC})$ and similar expressions for the two other pairs of factors. Hence, we get

$$E = -\delta(\overrightarrow{AB},\overrightarrow{AC})\,\delta(\overrightarrow{BA},\overrightarrow{BC})\,\delta(\overrightarrow{CB},\overrightarrow{CA}).$$

By using here Axiom III 2, we see that one of these factors is -1 and the two others are +1, which implies again $E = 1$.

Hence, we always have

$$\delta(\overrightarrow{AP},\overrightarrow{BQ})\delta(\overrightarrow{BQ},\overrightarrow{CR})\delta(\overrightarrow{CR},\overrightarrow{AP}) = 1.$$

It follows that $\delta(\overrightarrow{AP},\overrightarrow{BQ}) = -1$ iff the two remaining factors have different signs, i.e., iff the segments $\overrightarrow{AP},\overrightarrow{BQ}$ belong to different classes.

From (1), it is now obvious that condition a) holds, and from (3) that condition b) holds.

Hence, we have an existence proof as well, and because of the above uniqueness proof, the decomposition does not depend on the choice of $\overrightarrow{CR}$.

Theorem 5 is now completely proven.

The two classes of oriented segments, defined by Theorem 5 are unique and we may only change their notation. It is usual to give, in an arbitrary manner, to the segments of one of these classes the name of *positive segments* and to the segments of the second class the name of *negative segments*, denoted respectively as $\overrightarrow{AP} > 0$ and $\overrightarrow{BQ} < 0$.

We have now the possibility to define a very important binary relation on the set of points of a line, namely, we shall say that $A < B$ iff $\overrightarrow{AB} > 0$ and that $A \leq B$ if $A < B$ or $A = B$.

6. THEOREM. The relation $\leq$ is a total ordering relation on the line and it has the property that $A - B - C$ iff either $A < B < C$ or $C < B < A$.

Indeed, the relation $\leq$ is clearly reflexive. If $A \leq B$ and $B \leq A$ and if we had $\overrightarrow{AB} > 0$ and $\overrightarrow{BA} > 0$ we would contradict property b) of Theorem 5. The only remaining possibility is $A = B$ which shows that the relation $\leq$ is antisymmetric. Next, consider $A \leq B$ and $B \leq C$ with $\overrightarrow{AB} > 0$ and $\overrightarrow{BC} > 0$. Then $\overrightarrow{BA} < 0$ and we have $A - B - C$. This implies $\overline{B - A - C}$, which shows that $\overrightarrow{AB}$ and $\overrightarrow{AC}$ are in the same class, hence $\overrightarrow{AC} > 0$ and $A \leq C$. Thus, $\leq$ is also transitive.

Finally, for any two points, either $A = B$, or $A \neq B$ and we must have one of the cases $\overrightarrow{AB} > 0$ or $\overrightarrow{AB} < 0$, i.e., we must have at least one of the relations $A \leq B$ or $B \leq A$, which proves the first part of the theorem.

As for the second part, if $A - B - C$, $\overrightarrow{BA}$ and $\overrightarrow{BC}$ belong to different classes, which implies $A < B < C$ or $C < B < A$. The converse assertion is also immediate.

7. DEFINITION. A total ordering relation of a line is said to be *compatible with the betweenness relation* if it has the property of Theorem 6.

8. THEOREM. On every line, there is a total ordering relation which is compatible with the betweenness relation, and this relation is unique up to going over to the inverse order.

In fact, the existence of this relation was proven by Theorem 6. Now, if we start with such a relation on a line, we may divide the oriented segments of the line into two classes by putting $\overrightarrow{AB} > 0$ if $A < B$ and $\overrightarrow{AB} < 0$ if $B < A$. It is obvious that the two classes satisfy property b) of Theorem 5. But, they also satisfy property a) of the same theorem because, if $\overrightarrow{AP}$ and $\overrightarrow{AQ}$ belong to different classes we must have either $P < A < Q$ or $Q < A < P$ and, in view of the compatibility property, we get $P - A - Q$. Similarly, we get the converse assertion: $P - A - Q$ implies that $\overrightarrow{AP}$ and $\overrightarrow{AQ}$ are in different classes. Hence, we have property a).

It follows that an order which is compatible with betweenness defines a decomposition of the oriented segments in classes as required by Theorem 5 and, because of the uniqueness of such a decomposition, we get that the compatible order is unique up to inversion or order, q.e.d.

We make now the convention that on the lines of an ordered space we choose a fixed compatible order, which we denote by $\leq$ and whose inverse order is denoted by $\geq$, and always, when speaking of an order on a line, take it to mean that fixed order. The existence of these orders explains the names of order axioms and ordered spaces.

By means of the order relation we may introduce a new geometrically important notion.

9. DEFINITION. Let d be a line and O a point of D. The sets $\{M/M \in d$ and $M < O\}$, $\{N/N \in d$ and $N > O\}$ are called the *half-lines of d with origin O.*

It is to be noted that, actually, we have but one type of half-lines and not two, because by going over to the inverse order, we go from one half-line to the other.

Generally, a point gives a decomposition of the line in two half-lines, but we are not sure that both of them are non-empty. Anyhow, it is clear that two points M and N belong to different half-lines iff $M - O - N$ and this property is characteristic for the decomposition of a line into two half-lines by means of a point O.

b. Plane and Spatial Order Properties

The last remark mentioned above suggests the idea of generalizing the considered decompositions to planes and to space. To do this we first prove

10. THEOREM. Let d be a line of some plane α. Then, the set of points of α which do not belong to d may be uniquely decomposed into two disjoint classes such that two points P and Q belong to different classes, iff the line d cuts the interior of the segment PQ. Similarly, if π is a plane, the set of all points of the space which do not belong to that plane may be uniquely divided into two disjoint classes such that two points P and Q belong to different classes iff the plane π cuts the interior of the segment PQ.

Note that we do not require that both classes be non-void.

To prove the first part of the theorem, let us consider some point $P \in \alpha$ and $P \bar{\in} d$. Next, put in class I those points Q of $\alpha - d$ such that d does not cut the interior of PQ, and in class II the points R such that d cuts the interior of PR (of course, if such points Q,R exist).

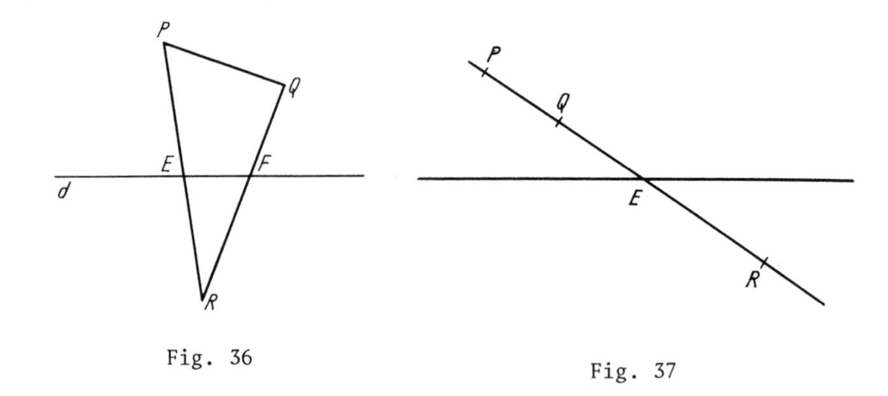

Fig. 36

Fig. 37

Now, suppose that Q and R are respectively in the classes I and II and that P,Q,R are noncollinear (Fig. 36). Then, we can use the Pasch Axiom for the traingle PQR and the line d, which cuts PR in the interior point E. We get that d cuts QR in the interior point F. Conversely, if QR is cut in an interior point, one of the segments PR,PQ is also cut in an interior point. Hence Q and R belong to different classes iff QR is cut by d in an interior point.

Next, if P,Q and R are as above but they are collinear (Fig. 37), then, by choosing an order on their line we have, for instance, $P < E < R$ and $P < Q < E$, which imply $Q < E < R$ and conversely (and similarly for all order possibilities of these points).

The first part of the theorem is thereby proved. As for the second part, the proof is exactly the same if using the intersection line of the planes π and PQR.

11. DEFINITION. The classes of points obtained by the above decomposition of a plane by a line are called *half planes* with the respective line as *origin*. The classes of points obtained by the decomposition of the space by a plane are called *half-spaces* with the respective plane as *origin*. If some geometric elements belong to different half-lines, half-planes or half-spaces respectively defined by a point, a line or a plane, we shall say that these elements *are on*

different sides (parts) of the corresponding origin. In the contrary case, they are *on the same side (part)*.

It is interesting to mention here that we may define now a topology of the space by taking the half-spaces as a subbasis of open sets. This is called the *ordered space topology* and it induces the plane topologies whose subbases are the half-planes and the line topologies whose subbases are the half-lines. These are respectively called the *topologies of the ordered planes or lines* (see §2 of the Introduction).

A simple application of Theorem 9 allows to give

12. DEFINITION. An *angle* is the geometric figure (i.e., set of points) consisting of two half-lines called *sides* which have the same origin called the *vertex* and which belong to different lines.

An angle is denoted by indicating the corresponding half-lines and their origin: $\widehat{hOk}$. The other classical notations for angles will be used as well.

An angle is obviously a plane geometric figure. Theorem 10 will allow us to define the interior and exterior of an angle:

13. DEFINITION. The point M of the plane α is an *interior point* of the angle $\widehat{hOk}$ situated in α if M is on the same side of the line Ok as the side h of the angle and, simultaneously, on the same side of the line Oh as Ok. A point which is neither interior nor on the sides of the angle is called an *exterior point*.

It follows that we have a decomposition of the set of points of the plane not situated on the sides of the angle into two classes: the *interior* and *exterior* of that angle. But we do not know whether both these classes are non-void.

14. PROPOSITION. If A and B are either two interior points of the angle $\widehat{hOk}$ or one is interior and one belongs to a side of $\widehat{hOk}$ or they are points belonging to different sides of this angle, and if M is a point such that $A - M - B$, then M is an interior point of $\widehat{hOk}$.

In fact, in the contrary case, we would have, for instance, A and M on different sides of h and, by taking the order $A < M$ the point $H = h \cap AM$ satisfies the relation $M < H < A$. Also, $A < M < B$. Or we would have $B > M > H > A$, in contradiction to the fact that B is interior to $\widehat{hOk}$. These contradictions imply the assertion.

15. PROPOSITION. Let be $\widehat{hOk}$, $A \in h$, $B \in k$ and $A - M - B$. Then the half-line OM is entirely in the interior of $\widehat{hOk}$. Conversely, if l is a non-empty half-line of origin O and interior to the angle $\widehat{hOk}$, l cuts every segment AB, joining points on different sides of that angle, in an interior point.

For the direct part of this theorem, if OM had some point in the exterior of the angle, OM would cut once again one of the sides of that angle. But then, OM is either h or k and $\overline{A - M - B}$ in contradiction to our hypothesis.

The converse part has a longer proof. Namely, besides O and B, the line Ok must have some third point C and we have three possible situations:

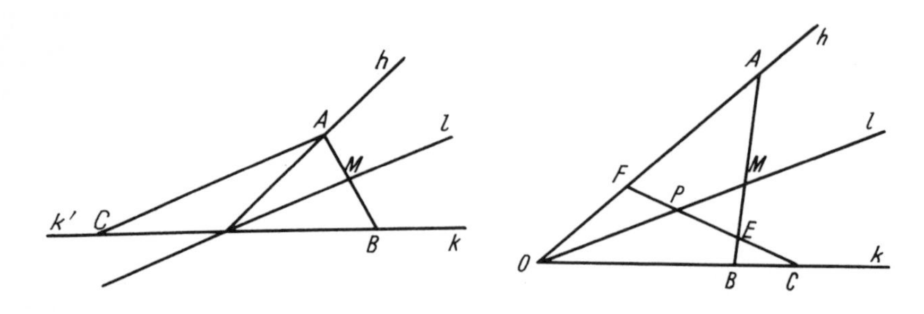

Fig. 38

Fig. 39

1) $C - O - B$ (Fig. 38). In this case, the result follows by using Pasch's Axiom for the triangle ABC and the line Ol.

2) $O - B - C$ (Fig. 39). In this case, let P be a point of l. If P and O are on different sides of AB, the existence of the required point is obvious. In the contrary case, C and P are on different sides of AB and there is a point $E = AP \cap CP$ with $A - E - B$. Using now the Pasch Axiom for the triangle AOB and the line CP, we get a point F such that $O - F - A$. Moreover, we clearly have $E - P - F$. Then, the existence of the point $M = l \cap AB$ is obtained by using Pasch's Axiom for the triangle AEF and the line l.

3) $O - C - B$ (Figs. 40, 41). In this case, taking P like in 2), CP cuts either OA (Fig. 40) or AB (Fig. 41) and the result follows easily by repeated applications of the Pasch Axiom. (The details are left to the reader to work out.)

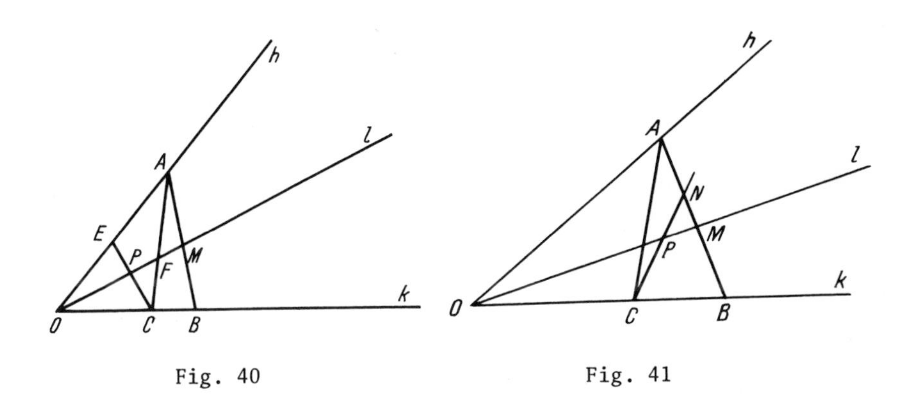

Fig. 40

Fig. 41

The notion of the interior of an angle may also be used in order to get some more important theorems about half-lines [3,10].

16. DEFINITION. The set of all half-lines with a given origin O and situated in a fixed plane α is called a *pencil of half-lines of origin O*. Such a pencil is denoted by $F(O)$. If $a,b,c,d \in F(O)$, we say that the pair (a,c) is *separating* the pair (b,d) if either a and c are collinear and b,d are

on different sides of their line or a and c define an angle and b,d are one interior and the other exterior to this angle.

One can show that this separating relation is symmetric with respect to the two pairs and that any four half-lines of $F(O)$ may be uniquely grouped into two separating pairs.

Next, if we fix some half-line $u \in F(O)$ and a,b,c are three other half-lines of the same pencil, we may define the relation $a - b - c$ (b lies *between* a and c) by the condition that the pair (u,b) separates the pair (a,b). This gives a *betweenness* relation in $F(O) - \{u\}$ and one can see that it has the properties of Axiom III 2 and Proposition 4.

All these proofs are by a straightforward analysis of all the possible cases and have a technical character. For instance, if we want to show that four half-lines may be grouped uniquely into two separating pairs, we shall take these half-lines a,b,c,d and fix three of them a,b,c. Suppose that these belong to different lines. Then either one of them is in the interior of the angle of the other two (Fig. 42) or none of them is such (Fig. 43). In each case, we study the regions into which the plane is divided and the possible position of d, and we get the desired result.

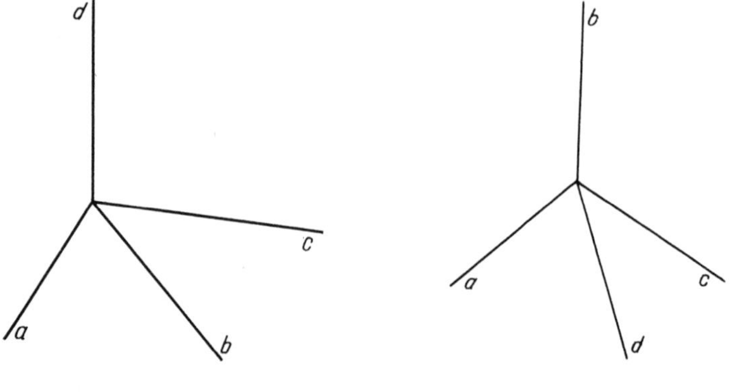

Fig. 42 Fig. 43

In a similar manner, the reader will be able to prove all the stated results.

Moreover, since these results are available, we may repeat for $F(O) - \{u\}$ all the necessary proofs performed previously on a line and derive Theorems 5, 6 and 8, i.e., the *linear order properties* for $F(O) - \{u\}$. As for the whole $F(O)$, one says that the separating relation for pairs defines on it a *cyclic order*.

The last problem to be discussed in this section is related to Theorem 5. This theorem belongs to a class of analogous results called *orientability properties*. The property given by Theorem 5 is known as the *orientability of the line*, and we say that the lines of an ordered space are *orientable* and they become *oriented* by choosing (arbitrarily) one of the corresponding classes of oriented segments as the class of the *positively oriented segments* (as we did while ordering the lines).

In an analogous manner, one may obtain the *orientability of the planes and of space*:

17. THEOREM. The set of ordered triangles of a given plane (respectively, the ordered tetrahedra of the space) may be uniquely decomposed into two disjoint non-empty classes, such that: a) a transposition of the vertices sends the triangle (tetrahedron) from one class to the other; b) if two triangles (tetrahedra) have a common side (face) they belong to different classes iff the remaining vertices are on different parts of the common side (face) and to the same class if those are on the same part of the common side (face).

Due to its similarity to the proof of Theorem 5 (using similar symbols δ) and its length, the proof of this theorem has been omitted. (This proof can be found in [15]). On the other hand, we shall study later on and by a different method, the orientability of ordered affine planes and spaces.

§2. POLYGONS AND POLYHEDRA

a. Convex Polygons

We have already encountered in the last section, some separation properties of the planes and space. Additional properties of this type appear in case of more general figures, namely for polygons and polyhedra. In this section, we intend to sketch a number of such separation properties.

From here on it will be convenient to call a *segment* not just a pair of points AB, but a set consisting of these points and all points lying between A and B, i.e., the *segment* AB will be understood as the corresponding closed interval on the line AB defined by using a compatible order on this line.

1. DEFINITION. If $A_0, \ldots, A_n$ are $(n+1)$-points with the property that any three consecutive points are noncollinear, the union of the segments A_0A_1, $A_1A_2, \ldots, A_{n-1}A_n$ is called a *polygonal line* with the *vertices* $A_0, \ldots, A_n$ and the *sides* A_iA_{i+1} ($i = 0, \ldots, n-1$). Such a line will be denoted by $A_0 \ldots A_n$. If $A_n = A_0$ (and, implicitly, $n \geq 3$), the polygonal line is called *closed* or a *polygon*, namely, for $n = 3$ - *a triangle*, for $n = 4$ - *a quadrangle*, etc.

   A polygon will be denoted by $A_1 \ldots A_n$. A polygon lying in a plane is called a *plane polygon* and *hereafter all the polygons will be plane, a fact which will not be mentioned specifically.*

   Two sides of a polygonal line are called *neighboring sides* if they are defined by three consecutive vertices; for a polygon the vertex A_1 is considered to succeed A_n.

   A polygonal line (polygon) is *simple* if two non-neighboring sides have no common points. A polygonal line (polygon) is *convex* if, with respect to every line of a side of the polygon, all the remaining sides of the polygon are in the same half-plane.

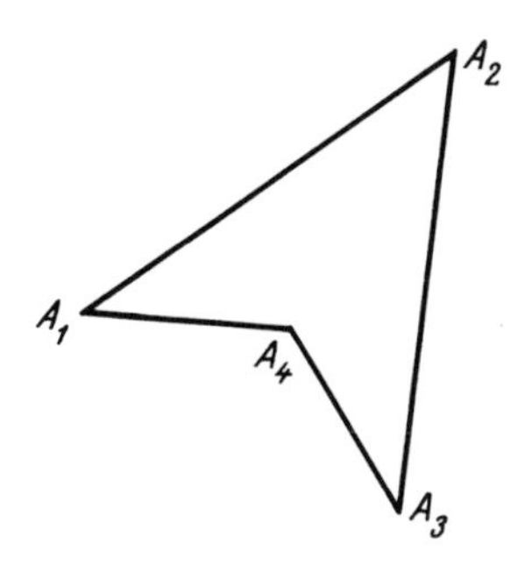

Fig. 44

We begin with the study of convex polygons. Obviously, the simplest polygon, i.e., the triangle, is convex and non-convex polygons must have at least four sides (Fig. 44).

2. PROPOSITION. A convex polygon is simple.

In fact, consider a non-simple polygon and suppose that, with the notation of Definition 1, A_1A_2 has common points with A_pA_{p+1} $(p \geq 3)$. If they have only one common point, which is either interior to one of them or a vertex, then the polygon cannot be in one half-plane with respect to both A_pA_{p+1} and A_1A_2. If A_1A_2 and A_pA_{p+1} have more than one common point, then, for instance, A_1A_2 and $A_{p+1}A_{p+2}$ will have only one common point and the conclusion follows like in the first case. Hence, a non-simple polygon is also non-convex which proves the proposition.

3. PROPOSITION. A line situated in the plane of a convex polygon and containing none of its sides has at most two points in common with the polygon.

Indeed, if the polygon is $A_1 \ldots A_n$ and the line is d and if they have three common points A, B, C (Fig. 45), and suppose we have $A - B - C$ with B on the line A_iA_{i+1} then we will have A and C on different sides of A_iA_{i+1}. This contradicts the definition of convexity and proves the proposition.

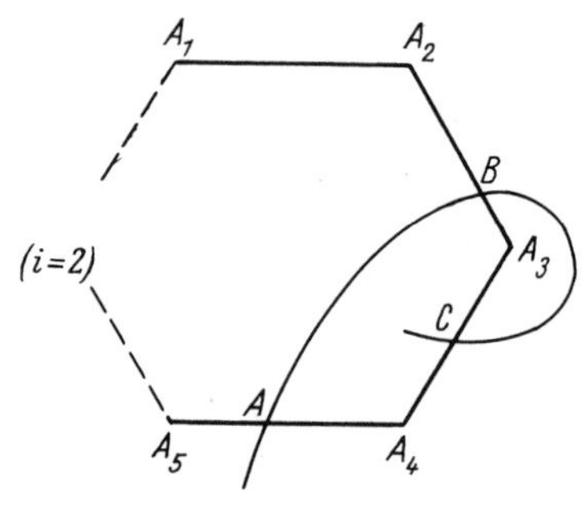

Fig. 45

Like angles, convex polygons define some subset of points of the plane which is called the *interior* of the polygon (We do not know however if this subset is non-empty.)

4. DEFINTION. Let $A_i A_{i+1}$ be a side of the convex polygon. The half-plane defined by $A_i A_{i+1}$ and containing the rest of the polygon is called the *positive half-plane* of that side, while the other half-plane is called *negative*. The *interior* of a convex polygon is the intersection of all the positive half-planes of its sides. The *exterior* of the polygon is the set of all the points of the plane which are neither interior nor on the polygon. Correspondingly, the points of the plane may be *interior*, *exterior* or on the polygon.

We can now prove

5. THEOREM. A segment joining two interior points of a convex polygon is entirely interior to the polygon. Every polygonal line which joins an interior and exterior point meets the polygon. Every two exterior points may be joined by a totally noninterior polygonal line.

The first assertion follows from the obvious fact that the points of the segment are in the positive half-plane of each of the sides of the polygon.

As for the second assertion, it is clear that the considered polygonal line has two neighboring vertices one interior and the second exterior with respect to the given polygon (if the second is on the polygon, there is nothing to prove), which means that it suffices to prove the result for the case when the polygonal line reduces to a segment. But then, we must have at least one side of the polygon such that the ends of the segment are in different half-planes of that side, which implies that the segment meets the corresponding line. If the intersection point is on the side of the polygon, we are done. If it is exterior, we take this point and the interior end of the segment and repeat the same reasoning. Since the polygon has a finite number of sides, we must finally obtain the required result.

Finally, for the last assertion, let P and Q be two points in the exterior of the convex polygon $A_1 \ldots A_n$ (Fig. 46) and let, for instance, P be on the negative half-plane of the side A_1A_2 on which we consider the order $A_1 < A_2$. Then, all the segments PA_i intersect A_1A_2 and we

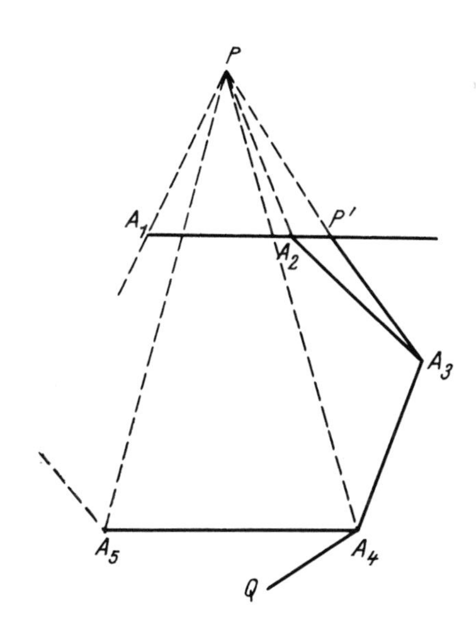

Fig. 46

have a greatest intersection point P'. If $P' > A_2$, we repeat this procedure with another line until we get a vertex of the polygon. Then, we proceed on the sides of the polygon until we reach some side with respect to which Q is in the negative half-plane and finally we joint our line to Q. If by going from P to P' and beyond we reach a side with Q in the negative half-plane, we have a completely exterior line. In any case our assertion is proven.

The reader is asked to analyse for himself similar assertions for the case when one or both of the points of Theorem 5 are on the polygon.

An immediate consequence of Theorem 5 is

6. COROLLARY. If P and Q are two points in the plane of a convex polygon and not on the polygon, and if every polygonal line which joins P and Q intersects the interior of the polygon, then either both points P and Q are interior or one of them is exterior to the polygon.

We also have the following result which completes Proposition 3.

7. PROPOSITION. Let P be a point either interior or on a convex polygon, and a line through P which contains no side of the polygon and is such that the polygon is not entirely in one half-plane of that line. Then d meets the polygon in exactly two points. If both of these points are different from P, P lies between them.

The proof is by induction on the number of the vertices of the polygon. Thus, for a triangle $\widehat{ABC}$, if P is an interior point (Fig. 47a), then P is interior for the angle ABC and, according to Proposition 1.15, BP intersects AC. Now, the required intersection points M,N of d with the triangle are defined by using the Pasch Axiom and the relation $M - P - N$ follows. If d contains A, B or C, or if P is on one of the sides of the triangle, the result is also obtainable by the Pasch Axiom.

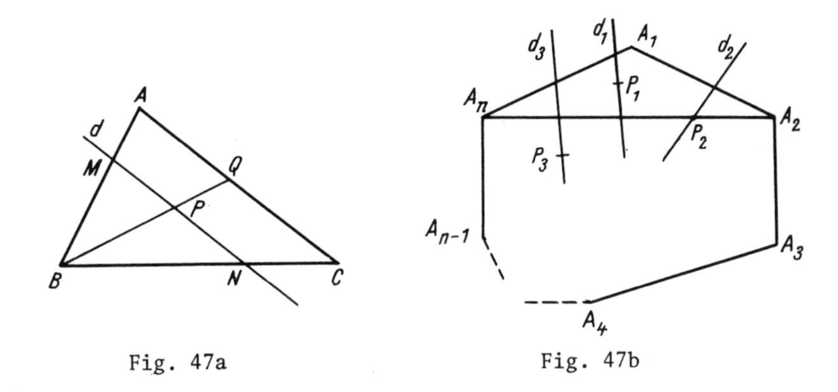

Fig. 47a

Fig. 47b

Suppose now that the proposition is true for polygons with n-1 vertices, then consider a polygon with n vertices (Fig. 47b). By joining A_n and A_2, we get the triangle $A_1A_2A_n$ and the polygon $A_2 \ldots A_{n-1}$ with n-1 vertices. Take the generic case where P is an interior point of the initial polygon. Clearly, it is in one of the positions P_1, P_2 or P_3 (Fig. 47b), and every time the desired result follows, by using for the line d, either the already proven case of a triangle or the induction hypothesis. The details are left to the reader to work out.

These considerations suggest the introduction of the following useful notions

8. DEFINITION. A set of points of the plane, which is the union of a convex polygon and of its interior is called a *convex plate*. A finite union of convex plates with disjoint interiors is called a *sum of convex plates*.

If p_i are convex plates with disjoint interiors, we denote by $s = \sum_i p_i$ their sum and we define its *boundary* $\dot{s}$ as the subset of the points lying on segments which belong to one and only one of the polygons of the plates p_i. Obviously for one plate p, the boundary is the corresponding polygon which, thus, must be denoted by $\dot{p}$. Our general convention for denoting the boundary of a sum will be

$$\dot{s} = \sum_i \dot{p}_i \pmod{2} ,$$

which is suggested by the fact that $\dot{s}$ is given by the disjoint union of the $\dot{p}_i$ in which the twice appearing segments (but not isolated points) are deleted. We note that a similar notational convention is available for writing a simple polygonal line as a sum modulo 2 of segments.

9. THEOREM. Let p be a convex plate and k_i $(i = 1,\dots,m)$ segments with ends on its boundary $\dot{p}$. Then p is decomposed into a sum of convex plates defined by portions of the segments k_i .

The proof is by induction on m.

Particularly, using the *diagonals*, defined as segments joining two non-consecutive vertices of the polygon, we may obtain a decomposition of any convex plate p into a sum of triangles.

b. The Jordan Separation Theorem

Now, consider again the situation described in Theorem 9 above, and denote by p_h the plates of the decomposition. Their boundaries consist of parts c_{h_k} either of the sides of p or of the segments k_i , such that we can write down

$$p = \sum p_h \quad , \quad \dot{p}_h = \cup c_{i_h} .$$

Next, if q is a convex plate defined in p by $s < m$ of the given segments $k_1,\dots,k_m$, the remaining segments define a decomposition of q into a sum of convex plates: $q = \sum q_a$, with the boundary $\dot{q} = \sum \dot{q}_a$ (mod. 2).

A more general result which holds for the same configuration is

10. LEMMA (Veblen). With the notation as above, every simple polygon $\dot{q}$ which is the union of some of the segments c_{h_k} may

be uniquely expressed as a sum $\sum \dot{q}_a$ (mod. 2), where the q_a are convex disjoint plates with disjoint interiors. Particularly, $\dot{p} = \sum_h \dot{p}_h$ (mod. 2).

The proof is by induction with respect to the number m of the segments k_i which contain sides of $\dot{q}$. For $m = 1$, the result clearly holds. Suppose it is true for less than m segments and let the sides of $\dot{q}$ belong to m segments q_i. Let $\dot{q} = \cup_{u=1}^{v} c_u$ (it is useless to put here two indices for the segments c) and let $c_v \subseteq k_m$. Then, there are just two plates p_h which have c_v as a side and we may take one of them, let's say q_1, such that $\dot{q} + \dot{q}_1$ (mod. 2) is a new simple polygon which does not have the side c_v anymore and whose sides belong to at most the same m segments k_i. Clearly, after some finite number of similar steps we get a simple polygon $\dot{q} + \dot{q}_1 + \ldots + \dot{q}_l$ (mod. 2) with no side on k_m and with all the sides on $k_1, \ldots, k_{m-1}$.

But then, we have by induction hypothesis

$$\dot{q} + \dot{q}_1 + \ldots + \dot{q}_l = \dot{q}_{l+1} + \ldots + \dot{q}_t \quad \text{(mod. 2)}.$$

Moreover (see, for instance) Fig. 49 later on) $q_1, \ldots, q_l$ may be chosen such that $\sum_1^t q_a$ is meaningful (which happens if we choose them such as to "cut down" parts of $\dot{q}$ and not to add). Then by adding, modulo 2, the terms $\dot{q}_1, \ldots, \dot{q}_l$ to the previous equality we get

$$\dot{q} = \sum_1^t \dot{q}_a \quad \text{(mod. 2)},$$

which proves the existence of the desired decomposition.

Now, if we also had $\dot{q} = \sum \dot{q}_b$ (mod. 2) and this is a different decomposition of $\dot{q}$, we would get, by equilizing the two decompositions, some relation of the form $\sum \dot{q}_c = 0$ (mod. 2), which is seen to be contradictory by a similar method, i.e., by induction on m and by eliminating k_m as above. Thereby we get the uniqueness of the required decomposition and prove the Veblen Lemma.

11. THEOREM. Let $A_1, \ldots, A_n$ be n coplanar points which are not collinear. Then, there is a unique convex polygon such that its vertices are some of the given points, and the other given points are either on the sides or in the interior of the polygon.

The proof of this theorem is by induction with respect to n and, since the points are non-collinear, $n \geq 3$. For $n = 3$, the required polygon is just the triangle defined by the three points.

Now, suppose the result holds for $n - 1$ points, then take n points $A_1, \ldots, A_n$.

Construct the convex polygon associated with the points $A_1, \ldots, A_{n-1}$ (or take the closed segment having two of them as ends and containing all the others in its interior, if these points are collinear).

If A_n is either on this polygon or in its interior, this is also the polygon of the n points.

If A_n is exterior, let it be, for instance, in the negative half-plane of A_1A_2 (Fig. 48). Then, for $i = 1, \ldots, n - 1$, A_nA_i intersect A_1A_2 and, by taking, for instance, the order $A_1 < A_2$, we have a smallest and a largest intersection point.

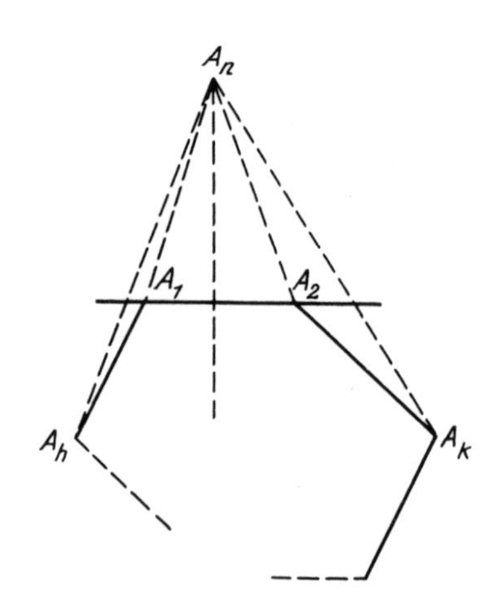

Fig. 48

There are two corresponding vertices A_h, A_k, and the convex polygon $A_h A_n A_k \ldots$, whose other vertices are those of the polygonal line joining A_h and A_k and not containing the segment $A_1 A_2$, is just the required polygon.

The uniqueness follows by a reductio ad absurdum, q.e.d.

12. DEFINITION. The convex polygon of Theorem 11 is called the *convex envelope* of the points $A_1, \ldots, A_n$.

Now, we have the possibility of studying simple polygons. The main result is the following famous theorem:

13. THEOREM (Jordan). If one has a simple polygon in a plane, the set of the points of that plane, not belonging to the polygon, may be uniquely divided into two disjoint parts called the interior and exterior of the polygon. Any polygonal line which joins an interior and an exterior point intersects the polygon.

Let p be our simple polygon and $\bar{p}$ the convex envelope of its vertices. Then, the prolongations of those sides of p which are in the interior of $\bar{p}$, are some segments k_i which give a decomposition like that of Theorem 9 (Fig. 49). Hence, we may apply Lemma 10 and,

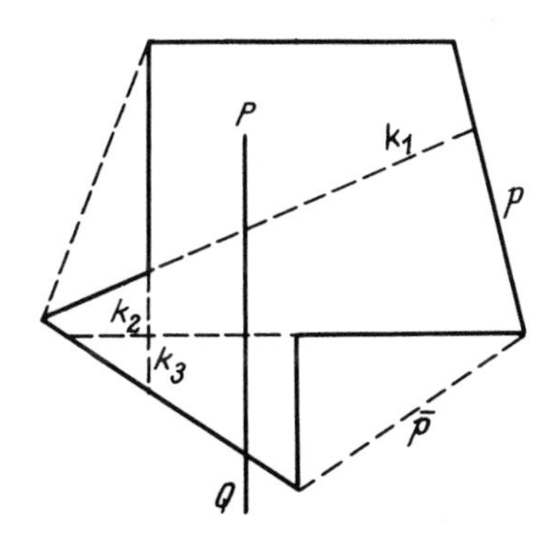

Fig. 49

with the notation there, we get some uniquely defined convex plates q_a such that $p = \sum \dot{q}_a$ (mod. 2), i.e., $p = \dot{s}$, where $s = \sum q_a$ is the corresponding sum of convex plates.

Consequently, we shall define the *interior* of p to be the set $s - p$ and the *exterior* of p to be the complementary set of s in the plane.

Further, for the last assertion of Theorem 13, we can remark that, just like in the case of the convex polygons (Theorem 5) it suffices to take only one segment joining the interior point P with the exterior point Q. But the segment PQ clearly has only a finite number of intersection points with sides of the polygons $\dot{q}_a$ above, and the last of them, in the order $P < Q$, must be on a side of P since, otherwise, there would be one more intersection point between that one and Q.

Jordan's theorem is thereby proven.

Now, results like Theorem 5, Corollary 6 and so on can be considered. One can also prove

14. THEOREM. Let p be some simple polygon and l a simple polygonal line which joins two points of p and is interior to p. Then l separates the interior of p into two disjoint parts.

The proof is by the same method as for Theorem 13.

Now, we can define *simple plates* and *sums of simple plates*. If a polygon is a sum of simple plates we say that these plates define a *polygonal subdivision* of the polygon. The plates are then called *faces* and their sides and vertices are the *edges* and *vertices* of the subdivision.

It is interesting that there is a fundamental relationship between these elements, as shown by

15. THEOREM (Euler). For every polygonal subdivision of a simple plate, which has v vertices, m edges and f faces, the following relation holds: $v - m + f = 1$.

In fact, let us begin by the remark that the desired relation holds for the trivial subdivision consisting just of the given polygon, because, in this case $f = 1$ and $v = m$. Now, consider subdivisions obtained from the trivial one only by adding vertices on the given polygon (this means that we now accept polygons with consecutive sides on the same line which does not matter here). Then again $f = 1$, and to v and m we add the same number h, such that the relation holds.

Consider a further intermediate subdivision obtained by introducing an interior polygonal line which joins two vertices on the given polygon. Such a line is called a *cut* (Fig. 50). Then we have one more face, α more edges and $\alpha - 1$ more vertices (the vertices on the given polygon are not new vertices), which yield the same value $v - m + f = 1$.

Finally, an arbitrary subdivision may be obtained, after inserting all its vertices on the given polygon, by a finite number of cuts which behave as above (Fig. 50).

Euler's theorem is thereby proven.

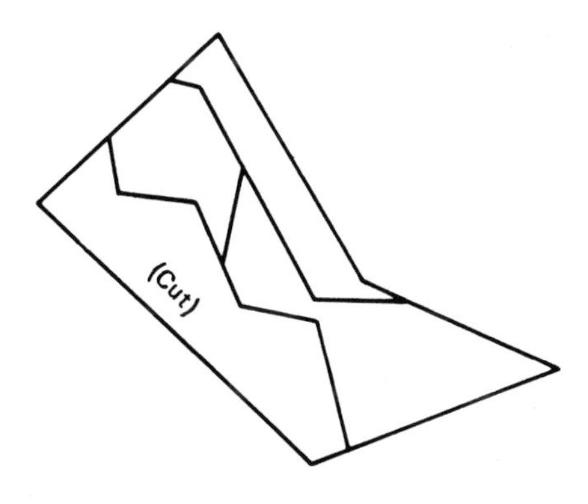

Fig. 50

c. Problems on Polyhedra

Now we shall outline some similar problems for the three-dimensional case. For a complete exposition, we refer to [10].

16. DEFINITION. By a *dihedral angle* we understand a pair of half-planes with the same origin and belonging to different planes. The line which is the common origin of the half-planes is called the *edge* of the angle and the half-planes are its *faces*.

Clearly, the interior and exterior of dihedral angles may be defined in the same manner as in the case of plane angles. We may define pencils of the half-planes, the separation relation for pairs of elements of a pencil, the cyclic order, etc. If we take a half-plane out of the pencil, we have a linear order for the rest.

17. DEFINITION. Consider a finite sequence of half-lines a_1, $\ldots, a_n$ with the same origin O and which are such that: a) any three successive half-lines are non-coplanar (a_1 has to be considered successive to a_n); b) two neighboring domains, defined by angles $\widehat{a_i a_{i+1}}$ together with their interiors, do not intersect each other. Such a sequence is called a *skeleton of a polyhedral angle* (n-hedral). O is called the *vertex*, a_i are called *edges* and the angles $\widehat{a_i a_{i+1}}$ with their interiors are the *faces* of this angle. The union of all the faces of such a skeleton is called a *polyhedral angle*. A polyhedral angle is *convex* if, for each of its faces, the rest of the angle lies in one half-space, called *positive*.

For polyhedral angles the following may be considered. The *interior* of a convex polyhedral angle may be defined as the intersection of all the positive half-spaces of its faces. This provides a decomposition of the set of points of the space, which do not belong to the polyhedral angle, into two disjoint parts, the interior and the exterior of the angle, with the same properties as in Theorem 5.

Next, the method which was used to prove the Veblen lemma and the Jordan theorem has a natural extension which leads to the definition of the interior and exterior of an arbitrary polyhedral angle (which is *simple* by Definition 17).

18. DEFINITION. Let us call a *face* any simple polygonal plate, while its sides and vertices will be called *edges* and *vertices* of the face. A *simple polyhedron* is a finite union of faces $f_1, f_2, \ldots, f_n$, which satisfy the following conditions: a) the intersection of two faces is either an edge or a vertex or the empty set; b) every edge belongs to exactly two faces; c) all the faces with a common vertex may be cyclically ordered such that two consecutive faces have a common edge; d) the above mentioned properties are not valid if we take out at least one face.

The simplest example is clearly a tetrahedron in which case the reader is asked to note how the properties a), b), c), d) appear. Another remark worth noting is that Definition 18 does not require that two faces with a common edge should be non-coplanar. If so desired, we may add this condition, which can be achieved by conveniently grouping the coplanar faces into one single face. If coplanar faces are admitted, then, by convenient decompositions, all the faces of a polyhedron may be considered triangles.

The following property is the *connectedness* of polyhedra.

19. PROPOSITION. Two arbitrary points of a simple polyhedron may be joined by a polygonal line situated on this polyhedron.

In fact, if the two points are on the same face or two faces with a common edge the result is obvious. As for the general case, we shall note that the faces which contain the given points may be joined by a chain of faces with common edges (*adjacent* faces). Indeed, if we start with the face of the first point and keep going to adjacent faces we shall have a finite chain containing all the faces of the polyhedra since, otherwise, we get a subsystem of faces satisfying the conditions a), b), c) of Definition 18, and this contradicts condition d) of the same definition. It is clear now that the two given points can be joined by a polygonal line which lies on the obtained chain of adjacent faces, and this proves the proposition.

20. DEFINITION. A simple polyhedron is called *convex* if for every face, the rest of the polyhedron is entirely in one half-space of that face, called the *positive half-space*. The intersection of all these positive half-spaces is called the *interior* of the convex polyhedron. The union of the polyhedron and of its interior is called a *convex body* and its complementary set is the *exterior* of the polyhedron.

Clearly, the faces of a convex polyhedron must be convex plates.

By means of Definition 20, we can generalize for convex polyhedra results like Proposition 3, Theorem 5, Proposition 7, Theorem 9, Lemma 10 and Theorems 11 and 13. Of course, we shall have also to study the intersection of a polyhedron with a plane. In the generalization of Theorem 9, we shall consider the decomposition of the polyhedron by plane sections and introduce sums of convex bodies and Σ (mod. 2) for their boundaries. This leads to the generalization of Veblen's lemma and of Jordan's theorem, which assures the existence of the interior of any simple polyhedron. As a consequence, every *simple polyhedral body* (i.e., a simple polyhedron and its interior) may be considered as a sum of tetrahedra. As for the generalization of Theorem 11, this gives the *convex envelope* of a finite set of spatial points.

We shall give here another important result.

21. THEOREM (Euler). If we denote by v the number of vertices, by m the number of edges and by f the number of faces of a convex polyhedron, the following relation holds: $v - m + f = 2$.

In fact, if we take out a face of the polyhedron, the remaining figure clearly behaves, from the viewpoint of Theorem 21, like a plane polygonal subdivision. Hence, by repeating the proof of Theorem 15, we get

$$v - m + (f-1) = 1,$$

which proves the theorem.

Actually, the Euler theorem can be extended to a larger class of simple polyhedra, namely to those for which every closed simple polygonal line on the polyhedron decomposes the polyhedron into two parts. Such polyhedra are called *simply connected* and the convex polyhedra are simply connected. This generalization of Theorem 21 has a similar proof. But Theorem 21 is not valid for non-simply connected polyhedra. For instance, for the polyhedron of Fig. 51, we have $v - m + f = 0$, but it is easily seen to be non-simply connected.

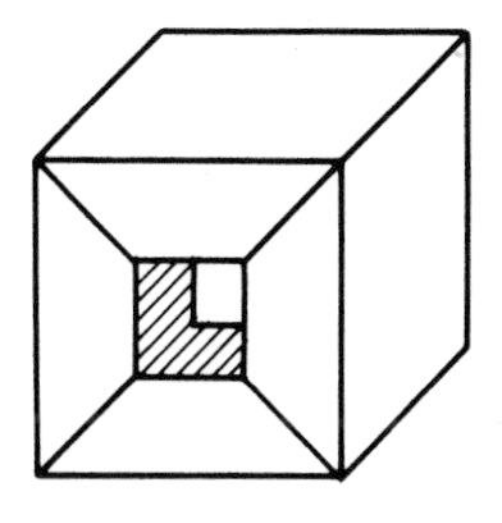

Fig. 51

We shall consider an interesting application of Theorem 21. Let us define a *quasi-regular polyhedron of type* (p,q) as a convex (or simply-connected) polyhedron such that all its faces have p vertices and through every vertex it has q faces. We want to determine all the possible types of quasi-regular polyhedra.

First, note that, in view of the existing cyclic order of the q faces of the same vertex, we have q edges through every vertex, whence, since an edge has two vertices, we get

$$m = \frac{qv}{2}.$$

Next, since there are q faces through every vertex and every face has p vertices, the number of faces is

$$f = \frac{qv}{p}.$$

Now, by Euler's theorem we have

$$v - m + f = v\left(1 - \frac{q}{2} + \frac{q}{p}\right) = 2,$$

which implies

$$1 - \frac{q}{2} + \frac{p}{q} > 0,$$

i.e.,

$$\frac{1}{p} + \frac{1}{q} > \frac{1}{2}.$$

But this last relation is easily seen to be equivalent to

$$(p-2)(q-2) < 4,$$

whence $p,q \leq 5$.

This fact and the previously established relations produce the following table of the possible quasi-regular polyhedra (note however that their existence is not yet proved)

p	q	f	m	v	Name
3	3	4	6	4	Tetrahedron
3	4	8	12	6	Octahedron
3	5	20	30	12	Icosahedron
4	3	6	12	8	Parallelepiped
5	3	12	30	20	Dodecahedron

For the regular polyhedra of Euclidean space, where the faces are equal regular polygons and their dihedral angles are equal, this table was known to the ancient Greek geometers, and the respective polyhedra are called the *Platonic polyhedra*.

We shall end this section with a few considerations on some figures which are more general than the simple polyhedral bodies and some related algebraic definitions which have a very important topological significance.

22. DEFINITION. A *simplex* of dimension 0,1,2,3 is, respectively, a point, a segment (together with the interior), a triangular plate, a tetrahedral body. A *simplicial complex* in three-dimensional space is a finite set K of simplices of different dimensions such that: a) if $s \in K$, all faces, edges and vertices of s also belong to K; b) any two simplices of K have either an empty intersection or a common face, edge or vertex. The union of the simplices of K is called a *triangulable figure* and the complex K is called a *triangulation* of this figure. The simplices of dimension 0,1,2,3 of the complex K are called respectively *vertices*, *edges*, *triangles* and *tetrahedra* of K.

For the following algebraic construction, we have to consider an *ordered complex* K, i.e., the simplicial complex K endowed with an order $A_1,\ldots,A_{N_0}$ of its vertices, whose number is denoted by N_0.

Let $\{s_\alpha^i\}$ ($\alpha = 0,1,2,3$; $i = 1,\ldots,N_\alpha$) be the set of the α-dimensional *ordered simplices* of K. For each α, we consider the free abelian group generated by $\{s_\alpha^i\}$ (see §2 of the Introduction). Its elements are called α-dimensional *chains* of K and this group is denoted by $C_\alpha(K)$.

Let $l \in C_\alpha(K)$. We define the chain dl of the dimension $\alpha - 1$, called the *boundary* of l, which for simplices takes the values

$$dA_i = 0 \ ,$$

$$d(A_i A_j) = A_j - A_i,$$
$$d(A_i A_j A_k) = (A_j A_k) - (A_i A_k) + (A_i A_j),$$
$$d(A_i A_j A_k A_h) = (A_j A_k A_h) - (A_i A_k A_h) + (A_i A_j A_h) - (A_i A_j A_k),$$

and for $l = \sum_i m_i s_a^i$ (m_i-integers) is given by

$$dl = \sum_i m_i ds_a^i .$$

It is obvious that $d : C_a(K) \to C_{a-1}(K)$ is a homomorphism of groups and the reader can easily verify that for every chain l we have $ddl = 0$.

A chain l is called a *cycle* if $dl = 0$ and a boundary if $l = dh$ for some other chain h. If we denote by $Z_a(K)$ the set of α-dimensional cycles and by $B_a(K)$ the set of α-dimensional boundaries, it follows that $Z_a(K)$ is a subgroup of $C_a(K)$ (namely the kernel of d) and, because of $d_\circ d = 0$, $B_a(K)$ is a subgroup of $Z_a(K)$ (namely the image of d).

Now, since these are abelian groups, we may define

$$H_a(K) = Z_a(K)/B_a(K) \qquad (\alpha = 0,1,2,3),$$

which are just the groups that we were looking for. They are called the *homology groups* of the complex K and play an important topological role. However, a more ample development is beyond our aim here.

§3. ORDERED AFFINE SPACES

a. Equivalent Order Axioms

We now want to examine those ordered spaces where the parallel axiom II also holds. Consequently, in this section we assume the Axioms I, I' (i.e., the strong incidence axioms), II and III. We shall start by proposing some useful equivalent axiom systems.

1. PROPOSITION. In any ordered affine space, the parallel projection preserves the betweenness relations.

First, by *parallel projection* we understand either a map $f: d \to d'$, between two coplanar lines, which is such that all the lines $Af(A)$ $(A \in d)$ are parallel (particularly, if d and d' are concurrent, their intersection point is fixed by f) or a similar map between skew lines d, d', which sends every point $A \in d$ to $f(A) \in d'$ defined by the intersection of d' with a plane through A and parallel to some fixed plane. In both cases, we suppose that $f(A)$ exists for some point A and then it is obvious that f is bijective.

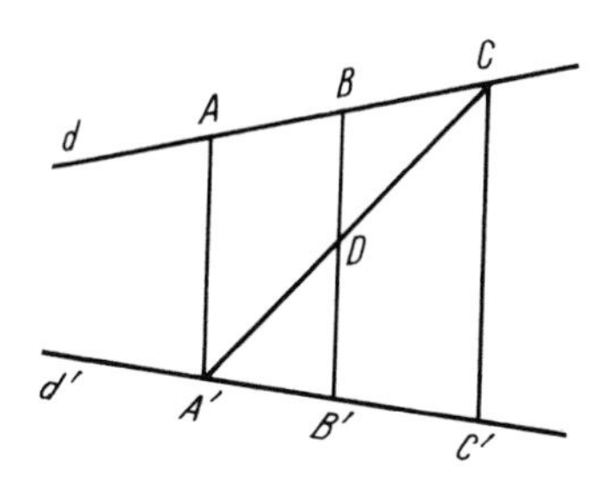

Fig. 52

Now, let d and d' be two coplanar lines, $A - B - C$ be three points of d and A', B', C' their images by some parallel projection (Fig. 52). Then, using the Pasch Axiom for the triangle $AA'C$ and the line BB' we get a point D such that $A' - D - C$, and, by the same axiom for the triangle $A'CC'$ and the line BB', we get $A' - B' - C'$. If, for instance, $B = B' = d \cap d'$ one has an obvious straightforward proof.

If d and d' are noncoplanar lines, the result follows similarly, using the same triangles and the intersection lines of their planes with the projection plane through B.

2. PROPOSITION. For any two points A and B, there is a point C such that $A - B - C$.

In fact, let d be a line which is parallel to AB and let $A' - B' - C'$ be three points on d. Project d onto AB by lines parallel to $B'B$ (Fig. 53). Let C and D be respectively the projections of C' and

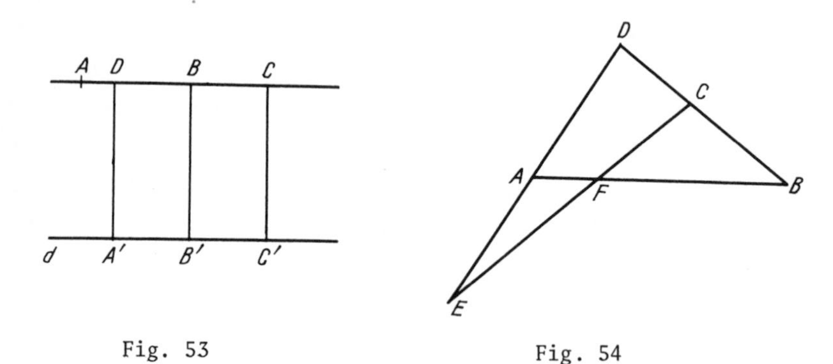

Fig. 53 Fig. 54

A' and suppose $\overline{A - B - C}$, i.e., either $A - D - B$ or $D - A - B$. Also, by Proposition 1, we have $D - B - C$. Then, if we choose on AB the order $A < B$ we must have either $A < D < B < C$ or $D < A < B < C$ and, in any case, $A - B - C$, q.e.d.

3. COROLLARY. For any segment, one has at least one interior point.

In fact, consider the segment AB and take a point C not on the line AB (Fig. 54). By Proposition 2 we also have a point D such that $B - C - D$ and next a point E such that $D - A - E$. Then, the Pasch Axiom applied to the triangle ABD and the line EC gives a point F such that $A - F - B$.

4. COROLLARY. Every line contains an infinite number of points. Every half-line, half-plane and half-space is non void. The interiors and exteriors of all simple polygons and polyhedra are non void.

In fact, it follows from Proposition 2 that the totally ordered set of the points of a line has neither a first nor a last element, i.e., it is infinite. This and Corollary 3 imply in a simple manner the other assertions of Corollary 4, and we leave the details for the reader to work out.

Actually, Proposition 2 provides us with many additional details about questions dealt with in the last two sections but which will not be treated here.

In Hilbert's original work [9], Proposition 2 was stated as one of the order axioms. More precisely, Hilbert's order axioms were III' 1 = III 1, III' 2 = Proposition 2, III' 4 = III 3 (Pasch) and

III' 3. If A,B,C are three collinear points, at most one of them is between the two others.

It follows, from the already established results that the axiom system (I, I', II, III) implies the system (I, II, III'). But the converse is also true. Actually, III' 2 assures the existence of three points on every line and the only remaining part is to prove III 2. But, if A,B,C are three collinear points, it follows from III' 3 that we have at most one of the relations $A-C-B$, $B-A-C$, $A-B-C$. If the first two of them do not hold, take a point D outside the line ABC and a point E such that $B-D-E$ (Fig. 55) and use Pasch's Axiom for the triangles BCE, ABE and the lines AD, CD. We get $C-F-E$ and $A-G-E$. Next, with the triangle CGE and the line AF we get $C-D-G$. Finally, with ACG and DE we get $A-B-C$ which is just III 2.

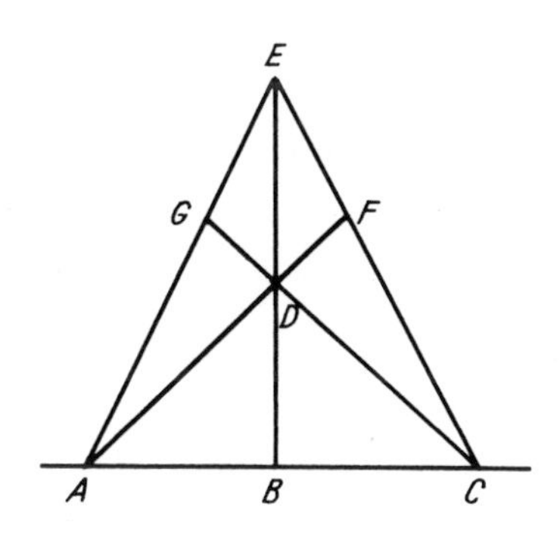

Fig. 55

Thereby we have proven

5. THEOREM. The axiom systems (I, I', II, III) and (I, II, III') are equivalent.

Another interesting formulation of the order axioms in an ordered affine space, due to E. Artin [1], consists of the following axioms:

III" 1. The set of the points of a line is endowed with a total ordering relation.

III" 2. The parallel projection between two coplanar lines either preserves or inverses the order.

We already proved (Theorem 1.8 and Proposition 1) that III" 1, III" 2 hold in an ordered affine space. Now, we shall show that, conversely, (I, II, III") implies (I, II, III).

Suppose (I, II, III") holds and define $A - B - C$ by the properties that these are three collinear points and, with respect to the order relation given on their line by III" 1, we have either $A < B < C$ or $A > B > C$. Then, the properties III 1 and III 2 are obviously satisfied.

In order to prove that III 3 is satisfied as well, we shall first show that the separation theorems into half-lines and half-planes are implied by the Artin axioms.

Indeed, let O be a point of a line d and define the half-lines with origin O as the subsets of the points of d which are, respectively, $< O$ and $> O$ for the order of III" 1. Then two points belong to different half-lines iff O is between them.

Further, let d be a line of a plane α and d' another, arbitrary, line of α which meets d at some point O. Denote by $p : \alpha \to d$ the map which *projects* all the points of α onto d', parallel to d (Fig. 56). Let d_1', d_2' be the two half-lines of d' with origin O. Then we have

$$\alpha - d = p^{-1}(d' - \{O\}) = p^{-1}(d_1') \cup p^{-1}(d_2') \quad ,$$

which is a well-defined decomposition of $\alpha - d$ into two disjoint

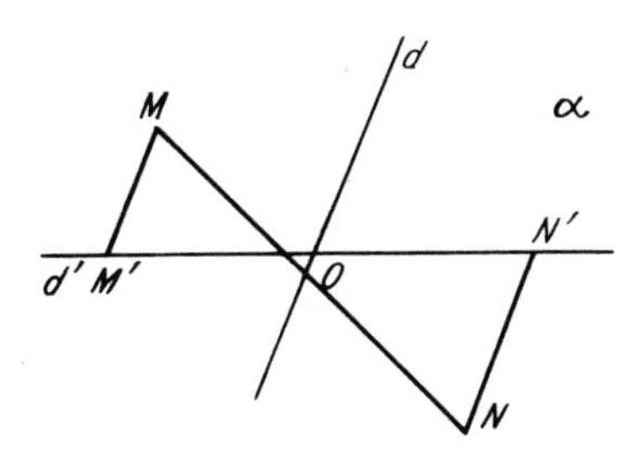

Fig. 56

classes. If M and N are points of different classes, $p(M) = M'$ and $p(N) = N'$ are on different half-lines and $M' - O - N'$, which, because of III" 2, implies that d cuts MN in an interior point. Conversely, if MN is cut internally by d, M' and N' belong to different half-lines and M,N belong to different classes. This shows that the obtained classes are just the half-planes defined by d in α.

It should be noted, however, that we do not know, at this point, whether all the half-lines and half-planes are non-void.

We are now able to verify the Pasch Axiom.

Consider an arbitrary triangle ABC and a line d in its plane, containing none of its vertices. Suppose also that d intersects AB at an interior point M. If, for instance, $d \parallel BC$, d intersects AC at an interior point because of III" 2 (Fig. 57) and similarly for $d \parallel AC$.

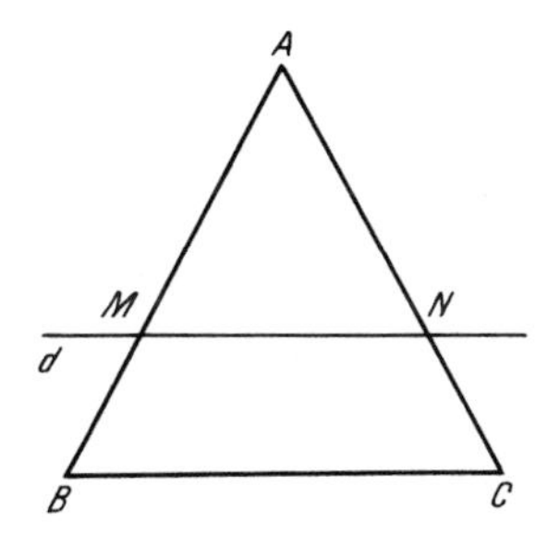

Fig. 57.

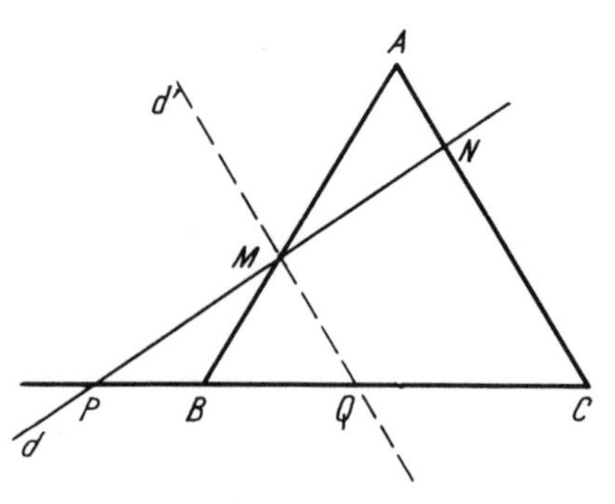

Fig. 58.

If d is not parallel with any of these lines and if d meets the line BC at a point P which is exterior to the segment BC (Fig. 58), we shall show that the intersection point N of d with AC lies between A and C.

Indeed, take $d' \parallel AC$ and $M' \in d'$. Then, by the already proven case, d' meets BC at Q such that $B - Q - C$, whence, for instance $P < B < Q < C$. It follows that the half-line MP and the line AC are in different half-planes with respect to d'. Hence, N is not on the half-line MP of d and it must be on the half-line AC with the origin A and on the half-line CA with the origin C. It follows $A - N - C$, q.e.d.

If we also accept the strong incidence axioms, we have proven

6. THEOREM. The axiom systems (I, I', II, III) and (I, I', II, III") are equivalent.

Hence, we may use as axioms of the ordered affine spaces the strong incidence axioms, the parallel axiom and Artin's axioms III". As in the case of the systems III and III', these axioms belong to plane geometry, while also defining the order properties of the whole three-dimensional space, (e.g., we may derive Proposition 1 showing that III" 2 is also valid for parallel projection by planes, etc.).

b. Determination of the Ordered Affine Spaces

In Chapter 1 we showed that every affine space is isomorphic to the affine space over a field K. Here, we want to obtain a corresponding result for the ordered affine spaces and, to do this, we begin by a study of the properties of the scalar field K of an ordered affine space.

We already know (Remark 1.3.16) that there is a bijective correspondence between the field K and the set of the points of a line and this suggests, of course, introducing a total ordering relation on K. Consider $k_i \in K$ $(i = 1,2)$, fix an origin O, a line d through O and a point $Q \neq O$ on d, take on d the order $O < Q$ and put $(P_i, Q)_O = k_i$. Now, we define $k_1 < k_2$ by the condition $P_1 < P_2$ (note that we have chosen $O < Q$).

The independence of this relation of the choice of d and Q is an immediate consequence of the definition of similar O-pairs (1.3.10) and of Axiom III" 2 and we leave the detailed proof for the reader. Also, using Axiom III" 2, one sees that when the origin O is changed and, hence, K is replaced by an isomorphic field K', then the isomorphism $K \approx K'$ of 1.3.15 preserves the order relation as well (i.e., it is a similarity of ordered sets). Indeed, this isomorphism is actually defined by a parallel projection. Hence we obtained

7. PROPOSITION. The coordinate (scalar) field of an ordered affine space has a canonical total ordering relation which makes it similar with the set of the points of a line.

Now, fields with a total ordering relation compatible, in a convenient sense, with the algebraic operations, are known in algebra, and are called *ordered fields*. To be more exact, we accept here the following definition:

8. DEFINITION. A field K together with a total ordering relation "<" on it is called an *ordered field* if: a) the mapping

$x \mapsto x+a$ where x is variable and a is constant in K is strictly monotonic; b) the mapping $x \mapsto xa$, where x and $a \neq 0$ are as above, either preserves or inverts the order.

At this stage, we can prove

9. THEOREM. A field K is the scalar field of an ordered affine space if and only if K is an ordered field.

In view of Proposition 7, the necessity part of this theorem follows if we prove the properties a) and b) of Definition 8. Condition b) is immediate: choose the origin O and put $a=(A,Q)_O$ $(A \neq O,\ Q \neq O)$, take next $x_1=(P_1,A)$, $x_2=(P_2,A)$ and suppose for instance that $P_1 < P_2$ with respect to the order $O<A$, which means $x_1 < x_2$; since $x_1 a = (P_1,Q)$ and $x_2 a = (P_2,Q)$, we see that, whenever the orders $O<Q$ and $O<A$ coincide, we have $x_1 a < x_2 a$ and, whenever these orders are opposite, $x_1 a > x_2 a$, q.e.d.

The proof of condition a) is more complicated. We have to show that, if for two arbitrary scalars we have $x_1 < x_2$, we also have $x_1 + a < x_2 + a$. Putting $x_i = (P_i,Q)_O$ $(i=1,2)$ and $a=(A,Q)$, we have $P_1 < P_2$ for the order $O<Q$ and, according to the definition of the sum of the scalars, we have to show that, if P_i' are the ends of the vectors $\overrightarrow{OP_i} + \overrightarrow{OA}$ $(i=1,2)$, then $P_1' < P_2'$ (Fig. 59).

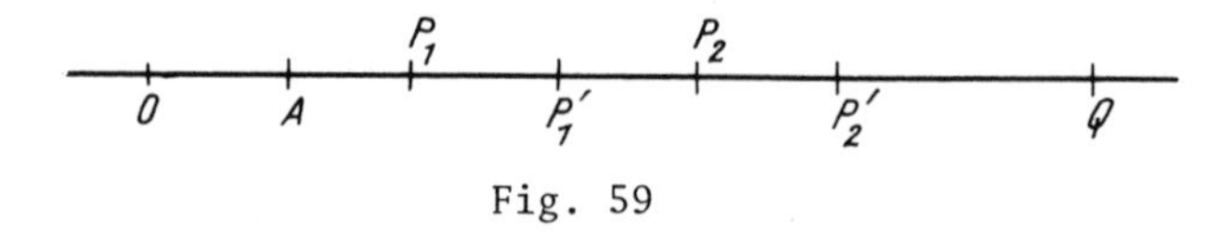

Fig. 59

Note that $\overrightarrow{P_1P_2} = \overrightarrow{P_1P_1'} + \overrightarrow{P_1'P_2'} + \overrightarrow{P_2'P_2} = \overrightarrow{OA} + \overrightarrow{P_1'P_2'} + \overrightarrow{AO} = \overrightarrow{P_1'P_2'}$. It follows that the desired result is equivalent to the fact that for any two oriented equipollent segments of an ordered line we have the same order between the origins and the ends.

Hence, let $\overrightarrow{P_1P_2} \sim \overrightarrow{P_1'P_2'}$ on a line d and let us express their equipollence by an intermediary segment $\overrightarrow{MN}$ on some line $d' \parallel d$ (Fig. 60),

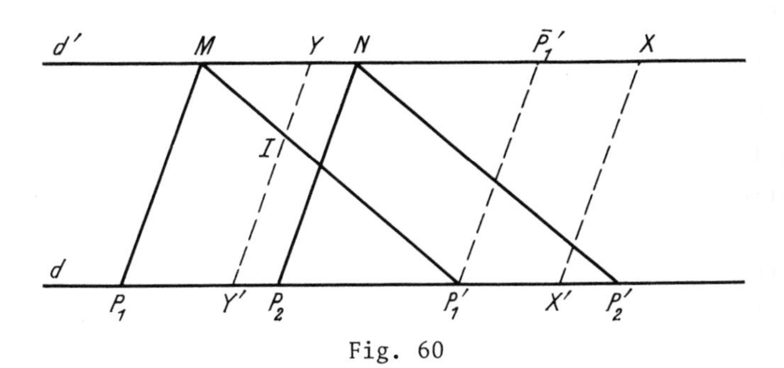

Fig. 60

i.e., $MP_1 \parallel NP_2$ and $MP_1' \parallel NP_2'$. We choose on d the order $P_1 < P_2$ and on d' the order $M < N$. Let us prove then that the points of d' which are $>M$ and the points of d which are $>P_1'$ are all in the same half-plane with respect to MP_1'.

Consider $\bar{P}_1'$ to be the intersection point of d' with the parallel through P_1' to MP_1 and, for instance, $P_1 < P_1'$. From the choice of the orders of d and d', it follows that the projection parallel to MP_1 preserves the order so that by III" 2 we have $M < P_1'$. Now, if $X > \bar{P}_1'$, $X \in d'$, we have also $X' > P_1'$ for $X' \in d$ and $XX' \parallel P_1M$ (in view of the above order preservation property) and X, X' are in the same half-plane with respect to MP_1'.

If, on the contrary, we take some Y such that $M < Y < \bar{P}_1'$, then take again $YY' \parallel P_1M$, $Y' \in d$ and use the Pasch Axiom for the triangle $MP_1'\bar{P}_1'$. We obtain some intersection point I of YY' and MP_1'. This point I is clearly "between" d and d' (which may be correctly expressed by using the half-planes with origins d, d'), which implies $Y - I - Y'$ and Y, Y' belong to different half-planes with respect to MP_1'. Since $Y' < P_1'$, Y belongs to the same half-plane of MP_1' as the points $>P_1'$ of d, and this is just the required result.

If $P_1' < P_1$, the proof is similar (we change the role of the lines d and d').

Further, it is immediate by now that the points of d' which are $>M$ and the points of d which are $<P_1'$ are in different half-planes with respect to MP_1'.

Then, if we had $P_2' < P_1'$ we would also get that NP_2' intersects MP_1' which contradicts the fact that these are parallel lines.

It follows that $P_1' < P_2'$ and the necessity part of Theorem 9 is thereby proven.

It is worth noting also that the last proof yields the following interesting result: *let d and d' be two parallel ordered lines; if there is one parallel projection between them which preserves (inverts) the order, then any other parallel projection between them behaves in the same manner.*

Let us now go over to the sufficiency part of Theorem 9. We take an arbitrary ordered field K and construct, as in 1.§4, the affine space over K. Then, we have to show that this space is a model for the Axioms III" 1, 2, too.

However, before proceeding, let us indicate a few other useful properties of the ordered fields.

Let K be an ordered field and $x \in K$. x is called *positive* if $x > 0$ and *negative* if $x < 0$, which (using a) of Definition 8) implies that $x < y$ if $y - x$ is positive or $x - y$ is negative. If x is positive, $-x$ is negative and conversely. We shall generally admit that the order of K is such that $0 < 1$ (otherwise we invert the order). By multiplying $0 < 1$ by $a \in K$ and using b) of Definition 8, we get that, if $a > 0$, the map $x \leftrightarrow xa$ preserves the order and if $a < 0$ this map inverts the order. Further, it follows from here that, if $x > 0$ and $y > 0$, then $xy > 0$ and, in a similar manner, we can obtain all the usual results for the sign of a product. Particularly, $xx^{-1} = 1 > 0$ shows that x and x^{-1} have always the same sign. For the sum, we have, for instance, that $x > 0$ and $y > 0$ imply $y < x + y$, whence $x + y > 0$, etc. Also, it is worth mentioning the following consequence of the last property: $1 + \ldots + 1 > 0$ for any number of terms, i.e., an ordered field K has, necessarily, the characteristic zero.

We come back now to the proof of the theorem.

Using analytic geometry in the affine space over K, we may represent parametrically a line as the locus of the points whose radius vectors with respect to a fixed origin are given by

$$\bar{r} = \bar{r}_0 + t\bar{v}, \tag{1}$$

where $\bar{r}_0$ defines a fixed point of the line, $\bar{v}$ is the direction vector and t is a parameter which runs over K.

Then, we get a total ordering relation on the line if we admit that the order of the points is the same as the order of the corresponding values of the parameter t.

This order is defined up to an inversion. Indeed, if the same line is given by the equation

$$\bar{r} = \bar{r}'_0 + t'(k\bar{v}) \tag{2}$$

we have, for two points $\bar{r}_i$ $(i = 1,2)$:

$$\bar{r}_0 + t_i\bar{v} = \bar{r}'_0 + t'_i k\bar{v} ,$$

whence

$$(t_1 - t'_1 k)\bar{v} = (t_2 - t'_2 k)\bar{v} = \bar{r}'_0 - \bar{r}_0 .$$

This implies

$$t_1 - t_2 = (t'_1 - t'_2)k,$$

and we see that the order of the values of t is either preserved, if $k > 0$, or inverted if $k < 0$. As for a change of the origin, it simply has no influence on t.

Hence, we see that Axiom III" 1 is satisfied.

Further, let

$$\bar{r} = \bar{a} + t\bar{v} \quad , \quad \bar{r} = \bar{b} + s\bar{w} \tag{3}$$

be two coplanar lines, $\bar{u} = h\bar{v} + k\bar{w}$ a vector belonging to their plane, and consider the projection between the two lines parallel with $\bar{u}$. Clearly, the relation between the values of the parameters of two corresponding points by the above projection is given by the condition that the vector defined by the two points be parallel to $\bar{u}$, i.e.,

$$\bar{b} + s\bar{w} - \bar{a} - t\bar{v} = p(h\bar{v} + k\bar{w}),$$

or if we consider that $\bar{a}$ and $\bar{b}$ are corresponding points of the considered projection (which is not a restriction of generality):

$$s\bar{w} - t\bar{v} = mh\bar{v} + mk\bar{w}. \tag{4}$$

Suppose now that $\bar{v}$ and $\bar{w}$ are linearly independent vectors. Then (4) implies $t = -mh$, $s = mk$, whence

$$s = -t(h^{-1}k). \tag{5}$$

This shows that the parallel projection preserves or inverts the order of the values of the parameters and also the order of the points on the given lines.

On the other hand, if $\bar{v}$ and $\bar{w}$ are collinear, we may assume them to be equal and, of course, we have to consider them noncollinear with $\bar{u}$. In this case, (4) holds iff $s = t$, whence the preservation of the order follows as well.

Hence, Axiom III" 2 is also satisfied and we finally have a complete proof of Theorem 9.

Moreover, an examination of the proofs of Proposition 7 and Theorem 9 shows that the isomorphism, given by analytic geometry, between a given affine space and the affine space over the scalar field of the given space (1, §4) is an isomorphism in what concerns the order properties too.

Consequently, we have determined all the ordered affine spaces up to an isomorphism. *Namely, these are just the affine spaces over ordered fields.* As a matter of fact, we have thereby established the noncontradiction of the axiom system I, II, III (and their equivalent formulations) in the sense that these are, for instance, as consistent as the arithmetic of the rational numbers (which, in turn, is as consistent as the system of the natural numbers). In fact, the field of the rationals with the usual order is an ordered field and the affine space over this field is a model of the mentioned axiom system.

Using Theorem 9, we may derive some more interesting results. Thus, from this theorem, together with Proposition 7 and Corollary 4, it follows that an ordered field has, necessarily, infinitely many elements, which is also a consequence of the already noted fact that such a field is of characteristic zero.

Since the characteristic is zero, it follows from Theorem 1.3. 19 that every segment AB has a midpoint M and, because, as we know, M is characterized by $\overrightarrow{AM} \sim \overrightarrow{MB}$, we may derive by arguments as in the proof of Theorem 9 that $A - M - B$, i.e., *the midpoint of a segment lies between the ends of that segment.*

Also, it is worth showing that the Pappus-Pascal Theorem is independent of Axioms III as well, which means that there are non-Papian ordered affine spaces (and Arguesian planes) too. To show this, we must prove the existence of a non-commutative ordered field. We do this now, following [1].

Let K be an arbitrary ordered field (which may or may not be commutative) and let σ be an automorphism of K. Denote $\sigma(a) = a^\sigma$, $a \in K$. Next, let k be the set of the formal power series of the form $\sum_{-\infty}^{+\infty} a_i t^i$, where $a_i \in K$, t is an indeterminate and at most a finite number of coefficients a_i with negative i are non-zero. Two such series are equal if their coefficients are equal for the same powers of t.

Define in k the sum as for polynomials and the product by

$$\left(\sum_i a_i t^i\right)\left(\sum_j b_j t^j\right) = \sum_r c_r t^r, \tag{6}$$

where

$$c_r = \sum_{i+j=r} a_i b_j^{\sigma^i}, \tag{7}$$

and σ^i is the i-th power of σ, i.e., $\sigma \circ \dots \circ \sigma$ (i times).

Then it follows easily that k becomes a field for which K is a subfield (if we identify t^0 with $1 \in K$). All the field properties, other than the existence of the inverse element, may be easily verified by the reader. The series 0 is the series with only vanishing coefficients and the series 1 has only one coefficient 1, for t^0, while all the other coefficients are 0.

As for the inverse, let's take $a = \sum_{i \geq n_0} a_i t^i$ $(a_{n_0} \neq 0)$ and try to define b_i such that the series $b = \sum_{i \geq m_0} b_i t^i$ $(b_{m_0} \neq 0)$ be such that $ab = 1$. By (7), we see that ab begins with $t^{n_0 + m_0}$, whence, necessarily, $m_0 = -n_0$ and a first coefficient identification gives

$$a_{n_0} b_{m_0}^{\sigma^{n_0}} = 1.$$

Since σ^{n_0} is an automorphism, this relation defines b_{m_0}. Analogously, by identifying the next coefficients in $ab = 1$, we define b_{m_0+1} and so on, the series b will be defined step by step.

Hence k is a field and we want to see whether k admits a suitable order.

If $0 \neq a \in k$, let us define as *the first coefficient* of a the first non-vanishing coefficient of the same series. Next, define $a > 0$ or $a < 0$ according to whether the first coefficient of a is positive or negative in K and put $a < b$ $(a, b \in k)$ if $b - a > 0$. Then, a trivial verification shows that "$<$" is a total ordering relation on k and satisfies property a) of Definition 8. As for property b) of the same Definition 8, it can be verified only if the automorphism σ has the property of sending positive elements to positive elements [1].

Hence, if σ preserves the set of positive elements of K, k is an ordered field and we shall also denote it by $k(K,\sigma)$.

Note now that from (6) and (7) we get

$$ta = a^\sigma t \qquad (a \in K),$$

whence we see that if σ is not the identical automorphism of K, k is non-commutative even if K is such.

Now, take the field Q of rational numbers and define first $K = k(Q, \text{id.})$ (we already know the id. - the identity - is the only automorphism of Q). This K is, by the above construction, a commutative ordered field. Next, the operation consisting of the replacement of the indeterminate t by $2t$ everywhere in the series of K, defines a nonidentical automorphism σ of K sending positive elements to positive elements. Using once again the above construction, we finally get an ordered non-commutative field $k(K,\sigma)$ as required and prove thereby the existence of ordered non-Pappian affine geometries.

c. Orientation of Ordered Affine Spaces

Finally, using Theorem 9, we shall study the orientability of ordered affine spaces. To do this, we shall need the theory of determinants over an arbitrary, generally non-commutative, field, as developed by J. Dieudonné. In the case of a commutative field, this theory reduces itself to the classical one.

Let K be an arbitrary field and K^* the multiplicative group of its non-zero elements. Also, let $\bar{K}^*$ be the abelian group obtained by factorizing K^* by its commutator subgroup (see §2 of the Introduction) and $\bar{K} = \{0\} \cup \bar{K}^*$. Generally, for $a \in K^*$ we shall denote by $\bar{a}$ the corresponding class of $\bar{K}$.

Now, we want to associate with every square matrix A with entries in K some elements of $\bar{K}$, called the *determinant* of A and denoted by det A or by $|A|$, such that the following conditions be satisfied:

1. If A' is the matrix obtained from A by the left-multiplication of a line by an element $k \in K$ then

$$\det A' = \bar{k} \det A.$$

2. If A' is obtained from A by adding one line to another, then

$$\det A' = \det A.$$

3. If I is the usual unit matrix, then

$$\det I = \bar{1}.$$

(In the sequel we shall denote $\bar{1} = 1$).

4. For all square matrices A and B we have

$$\det \begin{pmatrix} A & 0 \\ 0 & B \end{pmatrix} = \det A \cdot \det B,$$

where 0 denotes a block of zeroes.

Before proving that this is actually possible, let us derive a few consequences of the properties 1., 2., and 3. above.

a) *If we add to one of the lines of a matrix another line multiplied at the left with a factor of* K, *the value of the determinant remains unchanged.*

In fact, let us denote by A_i the lines of the matrix A, and start with the equality

$$\begin{vmatrix} \cdots\cdots \\ A_i + kA_j \\ \cdots\cdots \\ -kA_j \\ \cdots\cdots \end{vmatrix} = \begin{vmatrix} \cdots\cdots \\ A_i \\ \cdots\cdots \\ -kA_j \\ \cdots\cdots \end{vmatrix}$$

which is an easy consequence of 2. By 1., we may take out the factor $-\bar{k}$ and simplify it, which gives the desired equality (the case $k = 0$ is trivial).

b) *If A is a degenerate matrix,* $\det A = 0$.

In fact, if you recall that a *degenerate matrix* is defined as a matrix which has no inverse matrix, then by classical linear algebra, this happens iff the lines of the matrix are linearly dependent. But in this case, one of the lines is a linear combination of the others and substracting from it this linear combination we get, by repeated use of a), that det A equals a determinant with a zero line. Hence, by 1., det $A = 0$, q.e.d.

c) *Any transposition of two lines of a determinant implies its multiplication by* $-\bar{1}$.

Indeed, we have

$$\begin{vmatrix} \cdots\cdots \\ A_i \\ \cdots\cdots \\ A_j \\ \cdots\cdots \end{vmatrix} = \begin{vmatrix} \cdots\cdots \\ A_i + A_j \\ \cdots\cdots \\ A_j \\ \cdots\cdots \end{vmatrix} = \begin{vmatrix} \cdots\cdots \\ A_i + A_j \\ \cdots\cdots \\ -A_i \\ \cdots\cdots \end{vmatrix} = \begin{vmatrix} \cdots\cdots \\ A_j \\ \cdots\cdots \\ -A_i \\ \cdots\cdots \end{vmatrix} = -\bar{1} \begin{vmatrix} \cdots\cdots \\ A_j \\ \cdots\cdots \\ A_i \\ \cdots\cdots \end{vmatrix}.$$

We may now begin the proof of the existence and uniqueness of the determinants. The proof will be by induction on the order n of the matrix A.

Consider $n = 1$. By 3., we must have det $(1) = \bar{1}$ and by 1., we must have det $(a) = \bar{a}$ det $(1) = \bar{a}$, which shows that the determinants of order 1 are uniquely defined. Next, if we define det $(a) = \bar{a}$, 1., 2., 3., 4., are clearly satisfied. This proves the existence of the determinants of order 1.

Admitting now the existence and uniqueness of all determinants of order $<n$, let us show first the uniqueness of the determinants of order n.

We shall write a determinant of order n in one of the following ways, where the notation is obvious:

$$|A| = \begin{vmatrix} a_{11} \ldots a_{1n} \\ \cdots\cdots\cdots \\ a_{n1} \ldots a_{nn} \end{vmatrix} = \begin{vmatrix} A_1 \\ \cdots\cdot \\ A_n \end{vmatrix} = \begin{vmatrix} a_{11} B_1 \\ \cdots\cdots \\ a_{n1} B_n \end{vmatrix}.$$

If A is degenerate, we have $\det A = 0$. In the contrary case, A has an inverse matrix, whence it follows that there is a system of elements $k_i \in K$ ($i = 1,\ldots,n$), uniquely defined and not all zero, such that $\sum_{i=1}^{n} k_i A_i = (1,0,\ldots,0)$ and, implicitly, $\sum_{i=1}^{n} k_i B_i = 0$.

Take an index i such that $k_i \neq 0$. By the definition of the coefficients k_i and using classical transformations of determinants, but with the present properties, we get

$$(-\bar{1})^{i+1} \bar{k}_i \det A = \begin{vmatrix} 1 & 0\ldots0 \\ 0 & \\ \vdots & C_i \\ 0 & \end{vmatrix}$$

where C_i is a matrix of order $n-1$. Then, condition 4. implies

$$\det A = (-\bar{1})^{i+1} \bar{k}_i^{-1} \det C_i . \tag{8}$$

Now, if we also have $k_j \neq 0$, we get another expression of this type and we have to show that we actually get an equal value.

In the above notation, the lines of the matrix C_i are $B_1, \ldots, \hat{B}_i, \ldots, B_n$, where the sign "^" indicates the absence of the corresponding line. Denote by D the matrix obtained from C_i by replacing the line B_j by $k_j B_j$, and by E the matrix obtained from C_i by replacing the line B_j by B_i. Then, it follows

$$\det C_i = \bar{k}_j^{-1} \det D,$$

and next, using $\sum_{h=1}^{n} k_h B_h = 0$,

$$\det C_i = \bar{k}_j^{-1} (-\bar{k}_i) \det E.$$

But, by $|i-j|-1$ transpositions of lines the matrix E becomes C_j. Hence

$$\det C_i = (-\bar{1})^{i-j-1} \bar{k}_j^{-1} (-\bar{k}_i) \det C_j$$

and, because $\bar{K}$ is commutative (and this shows the necessity of the performed factorization by the commutator), we get

$$(-\bar{1})^{i+1}\, \bar{k}_i^{-1} \det C_i = (-\bar{1})^{j+1}\, \bar{k}_j^{-1} \det C_j \,.$$

Hence, if the determinants of order n exist, they are uniquely defined.

Finally, in order to prove the existence of these determinants, we shall show that, if they are defined by (8), the conditions 1.-4. are satisfied.

1. Suppose that the i-th line of the matrix is multiplied by $k \neq 0$. Then C_i of (8) remains unchanged, but the coefficient k_i is replaced by $k_i k^{-1}$, whence the desired result is immediate. Also, for $k = 0$ the result is trivial.

2. Suppose that the line A_i is replaced by $A_i + A_j$. Then the relation

$$k_i\,(A_i + A_j) + (k_j - k_i)A_j = k_i A_i + k_j A_j$$

shows that all coefficients k_h with $h \neq j$ remain unchanged and k_j has to be replaced by $k_j - k_i$. Now, if $k_h \neq 0$ with $h \neq j$ exists, the formula (8) with $i = h$ is not at all modified. If such k_h does not exist, only $k_j \neq 0$, i.e., $k_j B_j = 0$ and $B_j = 0$ so that the matrix C_j is the same for the matrix A and for A'. Then, the result follows by using formula (8) where i is replaced by j.

3. This is obvious if we use the induction hypothesis.

4. Take

$$A = \begin{pmatrix} B & 0 \\ 0 & C \end{pmatrix}.$$

If either B or C is degenerate, then, because of the corresponding linear dependence of the lines, A is degenerate as well and $|A| = |B|\,|C|$ because both sides are zero.

For the general case, using the classical transformations of a determinant, we may bring B to the form

$$D(p) = \begin{pmatrix} 1 & 0\ldots0 & 0 \\ 0 & 1\ldots0 & 0 \\ \ldots & \ldots & \ldots \\ 0 & 0\ldots1 & 0 \\ 0 & 0\ldots0 & p \end{pmatrix}$$

such that $|D(p)| = |B|$. (If transpositions of lines are used, the corresponding sign is attached to p.) Also, the matrix C will be similarly transformed into

$$D(q) = \begin{pmatrix} 1 & 0\ldots0 & 0 \\ 0 & 1\ldots0 & 0 \\ \ldots & \ldots & \ldots \\ 0 & 0\ldots1 & 0 \\ 0 & 0\ldots0 & q \end{pmatrix}.$$

Then, we get det $B = \bar{p}$, det $C = \bar{q}$ and det $A = \overline{pq}$, which is just the desired property.

Therefore the existence of our determinants, over an arbitrary field, is proven.

Further, we are also interested in calculating the determinant of a product AB of two matrices of order n. We have already remarked that, by transformations which do not change the value of the determinant, we may take every matrix into a matrix of the form $D(p)$ mentioned above. (Note, that this shows that in the case of a commutative field the determinants above coincide with the classical ones.)

But the mentioned transformations are just the multiplication of the given matrix by some other, conveniently chosen, matrix, i.e., we may always put a given matrix A in the form $A = CD(p)$, which gives

$$AB = [CD(p)]B = C[D(p)B].$$

But, it is straightforward that $D(p)B$ is the matrix obtained from B by multiplication of the last line by p. If we also use once again the fact that the multiplication of any matrix by C does not change the value of the determinant, we have

$$\det(AB) = \det[D(p)B] = \bar{p} \det B = \det A \det B, \tag{9}$$

a result which we want to obtain. Particularly, it follows that $\det(A^{-1}) = 1/\det A$.

Suppose now that K is an ordered field. An arbitrary commutator $aba^{-1}b^{-1}$ is necessarily positive, by the established rules for the sign of a product. Because of the above, we may associate with $\bar{k} \in \bar{K}^*$ the sign of k, and this sign will depend on the class $\bar{k}$ only. Particularly, the determinants of the matrices with entries in K have a well defined sign which will be denoted by sign $|A|$.

Now, this long algebraic digression allows us to prove

10. THEOREM. The sets of vector bases on a line, in a plane and in the space may be uniquely divided into two disjoint non-empty classes such that two bases belong to the same class or to different classes according to whether the matrix of the transformation between the two bases has a positive or a negative determinant.

Indeed, it suffices to fix a basis B, and to put in class I all the bases which are obtained from B by a transformation with a positive determinant, and in class II all the bases which are obtained from B by a transformation with a negative determinant.

The two classes are clearly disjoint. The first contains at least B and the second contains the basis obtained by changing the sign of one of the vectors of B.

Next, if B_1 and B_2 are two arbitrary bases and $B_i = T_i B$ $(i = 1,2)$, where T_i are the corresponding transformations, we have $B_2 = T_2 T_1^{-1} B_1$, which, in view of the above result regarding the determinant of the product of two matrices, implies the assertion of the theorem.

The property of Theorem 10 is called *the orientability property* of the ordered affine lines, planes and spaces, and *orienting the space* means choosing arbitrarily one of the two classes of bases as the class of *positive bases*. Then, the other class is the class of *negative bases*.

Finally, let us note that the above theory of determinants is also very useful in studying affine transformations in arbitrary and in ordered affine spaces. Namely, it is used in the analysis of the structure of the affine group [1].

§4. CONTINUITY AXIOMS

a. The Dedekind Continuity Axiom

The most important affine geometry is the geometry associated with the field of the real numbers, which is an ordered field. We should like to characterize it in the set of ordered affine spaces by a new group of axioms. This is the purpose of the present section.

While we simply might add the axiom that "the scalar field of the space is the real field," it is however more interesting to obtain this as a result of some axioms of a geometrical nature, which also clarify some other aspects.

The ordered field of the real numbers is characterized by some special order properties so that it is natural to require new conditions for the order of the points on a line. Namely, these properties are contained in the so-called Dedekind Principle, which we shall define now.

1. DEFINITION. Let $(M,<)$ be a totally ordered set. This set is said to be *dense* iff for every $a,b \in M$, there is some $c \in M$ such that $a < c < b$. In this case, a decomposition $M = I \cup S$,

where I and S are disjoint nonempty subsets of M is called a *Dedekind cut*, with the *lower class* I and the *upper class* S, iff every element of the lower class is smaller than every element of the higher class. The cut will be denoted by (I,S).

The simplest example of a cut is given by

$$I = \{a/a \in M \text{ and } a \leq a_0\}, \quad S = \{b/b \in M \text{ and } b > a_0\},$$

if these are non empty and $a_0 \in M$. This is called the *cut defined by the element* a_0 and a_0 is the largest element of the lower class. If we move a_0 from I to S, we get a similar situation, but this time the upper class has a smallest element a_0. For the sake of simplicity, we shall identify these two cuts.

Conversely, a Dedekind cut is of the type mentioned above if either the lower class has a largest element or the upper class has a smallest element. It is impossible to have them both since then (because of the density property), between these two elements of M there must be another one, and $M \neq I \cup S$, which is a contradiction. Hence, in our case, we may always consider that we have a largest element in the lower class (after an eventual move of an element from S into I). A Dedekind cut which has such a largest element of I is called of *the first kind*. A Dedekind cut which is not of the first kind is said to be of the *second kind*.

2. DEFINITION. Let M be a totally ordered dense set. M is said to satisfy the *Dedekind Principle* if every Dedekind cut of M is of the first kind.

The typical examples related to these definitions are offered by the set Q of the rational numbers and the real field R. Q has Dedekind cuts of the second type. For instance,

$$I = \{q/q \in Q, \ q < 0 \text{ or } q^2 < 2\},$$

$$S = \{q/q \in Q, \ q > 0 \text{ and } q^2 > 2\},$$

define such a cut. If the elements of Q are identified with the cuts of the first kind which they define, and if we add the set of cuts of the second type of Q, we get just the set R of the real numbers, and this set satisfies the Dedekind Principle [7, 8]. Actually, this procedure may be extended such as to obtain an embedding of any totally ordered dense set into another set which satisfies Dedekind's Principle [7].

Because the lines of an ordered affine space are dense ordered sets, and they are similar to the scalar field of the space, it is but natural to take as a new axiom just the Dedekind Principle, which, in this case, has a geometric formulation. We get thus the required group of axioms, called the *continuity axioms* and consisting of

IV. Every Dedekind cut on a line is of the first type.

b. Continuously Ordered Spaces

Before going on with the consequences of Axiom IV in affine geometry, it is worth noting that the axiom is also meaningful in any geometrically ordered space in which the lines are dense with respect to the order given by Theorem 1.8. Since this last property follows from the order axioms III' of §3, we are now led to study the geometry of the spaces which satisfy Axioms I, III' and IV. Such spaces will be called *continuously ordered spaces*.

We begin by showing that Axiom IV is equivalent with some topological property, i.e., this axiom is related to the topological properties of the space.

We already know, from §1, that the ordering properties of the space allow defining the ordered space topology, which induces on every line a topology of the line where the half-lines form a subbasis. If axioms III' are accepted, it is easy to see (and we leave this as an exercise for the reader) that this topology of the line is characterized by the fact that the open intervals (segments without their ends) containing a given point form a basis of open neighborhoods of that point (Introduction, §2).

We should also remark that in a similar way the topology induced in a plane is characterized by the fact that the open triangles (or simple plates) containing a point form a basis of open neighborhoods of that point and, in the space, such bases are given by the interiors of tetrahedra (or of simple polyhedra) [3]. All further references to topological properties are for the topologies which we have discussed here.

We can prove now

3. THEOREM. If Axioms I and III' are accepted, then Axiom IV is equivalent with topological connectedness of every line.

In fact, suppose Axiom IV holds and consider that for the line d we have $d = M \cup N$, where M and N are closed non-void subsets of d. Then, we shall show that $M \cap N \neq \phi$, which implies the connectedness of d (§2 of the Introduction).

Take $A \in M$, $B \in N$ and $A < B$ (otherwise we invert the order of d). Define

$$I = \{P \in d / P < A \text{ or } A \leq P \leq B \text{ and such that}$$
$$A < Q < P \text{ implies } Q \in M\},$$

$$S = d - I.$$

Then I and S are disjoint subsets of d, I contains A and S contains the points $>B$ and, if $P \in I$, $R \in S$, there is at least one point of N lying between them, which implies $P < R$. Hence, (I,S) is a Dedekind cut and, by IV there is some point D, which defines this cut and is such that $A \leq D \leq B$.

First, if $A < D < B$ and $X < D < Y$ is any neighborhood of D contained in the open segment AB, then any relation $X < Z < D$ implies $Z \in I$ and a relation $X < U < Z$ implies $U \in M$. I.e., the arbitrary neighborhood XY

of D has points in M. Also, if $D < Z < Y$, $Z \in S$, and between D and Z we must have points of N. It follows that D is an adherence point for both M and N and since these are closed we get $M \cap N \neq \phi$, q.e.d.

If $D = A$, it suffices to show, similarly, that D is an adherence point of N and if $D = B$ that it is an adherence point for M.

It is worth noting that the proof remains valid for every dense ordered set, satisfying the Dedekind principle.

Conversely, suppose d is connected and let (I,S) be a Dedekind cut of d. Then, denoting by a bar the topological closure of a set, we have $d = \bar{I} \cup \bar{S}$ which implies $\bar{I} \cap \bar{S} \neq \phi$. A straightforward examination of neighborhoods, using the definition of a cut, shows that $\bar{I} \cap \bar{S}$ has just one point (the contrary hypothesis leads to a contradiction). Also, this common point has to be either the largest in the lower class or the smallest in the upper class because, otherwise, it contradicts the property of being an adhecence point to both I and S. Hence the cut (I,S) is of the first kind, which completes the proof of Theorem 3.

4. COROLLARY. A continuously ordered space is connected and every plane of such a space is connected as well.

Indeed, if we had a decomposition $M \cup N$ of the space, where M and N are closed disjoint and non-empty, then, taking points $A \in M$, $B \in N$ and denoting by d the line AB we would have

$$d = (d \cap M) \cup (d \cap N)$$

and d would not be connected, in contradiction to Theorem 3. The connectedness of the planes follows similarly.

Another related and important theorem is

5. THEOREM (Bolzano-Weierstrass). Every bounded infinite set of collinear points has at least one limit point.

In fact, let M be such a subset of d, which means that there is some segment AB, with $A < B$ and containing M. Define on d the subset I

of the points P which are either $\leq A$ or between A and B and such that at most a finite number of points of M are between A and P. Next, define $S \subseteq d$ as the subset of the points Q, with the property that $A < Q$ and between A and Q there are infinitely many points of M.

Then, $d = I \cup S$, I and S are nonempty and disjoint and between any point of I and any point of S there is at least one point of M.

Hence (I,S) is a Dedekind cut and we have the corresponding definition point D. Then, by examining an open segment which contains D, we see that it must have infinitely many points of M. It follows that D is a limit point of M, q.e.d.

6. COROLLARY. If we also require that the topology of a line has a countable basis, then every closed segment is compact.

This follows by standard point-set topology [4].

7. COROLLARY. Let $\{A_i B_i\}$ $(i = 1,2,\dots)$ be a sequence of collinear points such that $A_{i+1} B_{i+1} \subseteq A_i B_i$ and such that for every i there is an index $j(i)$ for which $A_{i+j} B_{i+j}$ is contained in the interior of $A_i B_i$. Then $\cap_i$ [interior $(A_i B_i)$] $\neq \phi$.

If a sequence of segments satisfies the hypotheses of this corollary, we say that it is a *sequence of strictly including segments*. Obviously, in this case $\{A_i\}$ is a bounded infinite set of collinear points and if we take a limit point X of this set we shall have $X \in \cap_i$ [interior $(A_i B_i)$] because relations like

$$A_{n-1} < X < A_n < B_n < B_{n-1} ,$$

$$A_{n-1} < A_n < B_n < X < B_{n-1}$$

contradict the property of X being a limit point.

The next important result is

8. THEOREM. The cardinal number of the set of the points of a line is greater than the countable.

Indeed, if this set were the sequence $A_1, A_2, \ldots$ (where, generally, the order of elements of this sequence has nothing to do with the geometric order on the line), then either all $A_i \bar{\in} A_1 A_2$ $(i \geq 3)$ or there are points between A_1 and A_2. In the first case, we contradict the density property for the geometrical order of the line. In the second case, let i_1 be the smallest index of points A_{i_1} which are between A_1 and A_2.

Repeat the reasoning with $A_{i_1} A_2$ and get here an interior point A_{i_2} and so on.

Then we clearly get a strictly including sequence of segments, which by Corollary 7 have some common point and this point is not included in the original sequence A_i.

This is again a contradiction which proves the theorem.

Theorem 8 suggests that the simplest continuously ordered space to be considered is the space where a line is the closure of some countable subset of it, i.e., in the usual topological terminology, every line is a separable topological space. Under such hypothesis we can prove

9. THEOREM. If the line is a separable topological space, it is similar (from the order viewpoint) with the set of the real numbers.

Let $M \subseteq d$ be countable and $d = \bar{M}$. By neighborhood considerations, it is easy to see that the totally ordered set M (with induced order) is dense and has neither a first nor a last element.

But one can prove that every countable ordered set which is dense and has neither a first nor a last element is similar to the ordered set Q of the rational numbers.

Indeed, it is known that Q is countable, whence we may write down

$$Q = \{q_1, q_2, \ldots\},$$

$$M = \{A_1, A_2, \ldots\}.$$

Then, we define a bijective correspondence $\varphi : Q \to M$ in the following way: Put $\varphi(q_1) = A_1$ and $\varphi(q_2) = A_{i_2}$, where A_{i_2} is the first element of the sequence M which has the same order relation to A_1 as q_2 to q_1. Then, let i_3 be the smallest index $\neq 1$, i_2 and put $\varphi^{-1}(A_{i_3}) = q_{j_3}$, where q_{j_3} is the first element of the sequence Q which has the same order relations to q_1 and q_2 as A_{i_3} to A_1, A_{i_2} and so on.

We get alternatively a value of φ and a value of φ^{-1}, which is always possible in view of the mentioned order properties of M. Clearly, in this way we obtain the bijective correspondence φ which is a similarity of Q and M.

Further, an arbitrary real number may be identified with a Dedekind cut in Q which obviously has a corresponding (by φ) Dedekind cut in M. But this defines a Dedekind cut on d (an exercise for the reader) and a corresponding definition point which will be considered as the image of the real number we started with.

The condition $d = \bar{M}$ allows us to define similarly the inverse of this correspondence. Hence φ extends to a similarity $\bar{\varphi} : R \to d$, which completes the proof of Theorem 9.

10. COROLLARY. Under the hypotheses of Theorem 9, the lines are homeomorphic to the real field R endowed with the usual topology.

The proof is left as an exercise for the reader.

Finally, we want to obtain another interesting geometric property of the continuous ordered spaces, namely the existence of parallel lines. First, we consider

11. LEMMA. Suppose that the half-lines of some pencil are classified into two classes A and B such that: 1° the two classes are disjoint and each of them contains at least two elements, 2° if $a_1, a_2 \in A$ and $b_1, b_2 \in B$, the pairs (a_1, a_2) and (b_1, b_2) are not separating pairs. Then there are two half-lines (d_1, d_2) in this pencil such that (d_1, d_2) separates every pair (m, n) with $m \in A$ and $n \in B$.

Here, the separation relation is the one described in §1.

Let O be the centre of the pencil and p_i $(i = 1,2,3)$ half-lines such that $p_1 \in A$, $p_2 \in B$ and the prolongation of p_3 is interior to $\widehat{p_1 p_2}$; the proof of the existence of such p_i is left to the reader to work out. Then, consider $P_i \in p_i$ three arbitrary points and take first the line $P_1 P_2$ (Fig. 61), with the order $P_1 < P_2$. Define on this line the classes $I = \{Q/Q \leq P_1$ or $Q > P_1$ and $OQ \in A\}$ and $S = \{R/R \geq P_2$ or $R < P_2$ and $OR \in B\}$.

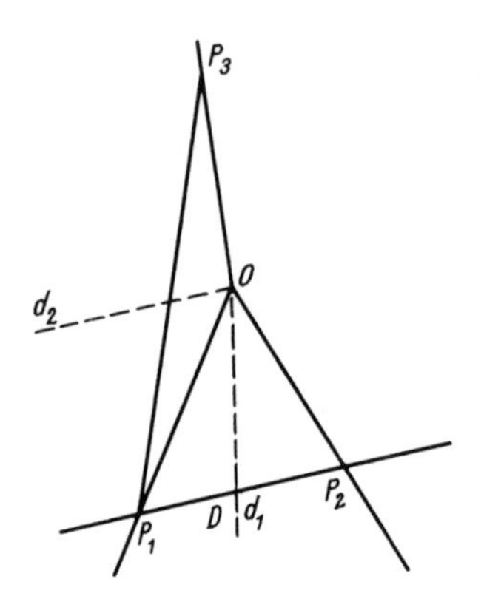

Fig. 61

Because of condition 2° of the lemma, we see that (I,S) is a Dedekind cut and has a corresponding definition point D $(P_1 \leq D \leq P_2)$ which gives the half-line $d_1 = OD$.

Then, if $p_3 \in A$, we work similarly with $P_2 P_3$, and if $p_3 \in B$, we work with $P_1 P_3$. Thus, we get a second half line d_2 and, by an analysis of all the possible positions, it follows that d_1 and d_2 are just the half-lines searched for.

Using the lemma, we shall prove [14]:

12. THEOREM. In every continuously ordered space, if a is a line and A a point not on this line, there is at least a parallel through A to a.

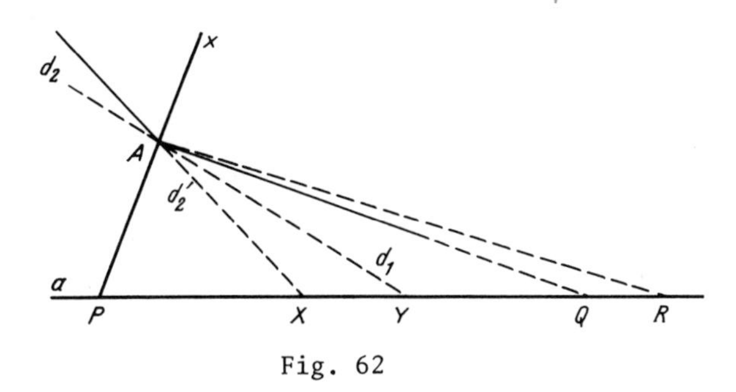

Fig. 62

Consider the mentioned line a and point A (Fig. 62). Divide the set of the half-lines with origin A and in the plane (A,a) into two classes, by putting in the first class the half-lines joining A to some point of a and in the second class all the others. Conditions 1° and 2° of Lemma 11 are then trivially verified, whence we may distinguish two half-lines of our pencil, d_1 and d_2 which separate the two classes. If P is an arbitrary point of a, then d_1, d_2 must be on different sides with respect to the line APx. Now, if $d_1 \cap a = Q$, take R such that $P - Q - R$, and note that AR and Ax belong to different classes. However, they are not separated by (d_1, d_2). This contradiction shows that $d_1 \cap a = \phi$. Similarly, $d_2 \cap a = \phi$. Admit now that the prolongation d_2' of d_2 cuts a in X and take $P - X - Y$. Then, the prolongation of AY and d_2' are in different classes although they are not separated by (d_1, d_2). This contradiction implies $d_2' \cap a = \phi$ and similarly $d_1' \cap a = \phi$.

Hence d_1 and d_2 define lines which are parallel to a and with it Theorem 12 is established. Note that, possibly, d_1 and d_2 are on the same line.

Hence (see 1, §2) as a consequence of adding the continuity axioms, our system is adequate to the geometries of Euclid and Lobachevsky, but the axiomatic construction of the Riemann geometry must take another road.

Since our interest is in Euclidean geometry, we admit I, III', and IV, as well as thereafter, II, and we shall study now the *affine continuously ordered spaces*.

We begin by remarking that, as a consequence of Theorem 12, Axion II may be replaced by a weaker axiom:

II' Through any exterior point of a line we have at least one parallel to that line.

In fact, Theorem 12 and Axiom II' imply Axiom II.

Next, from Proposition III.7 and Axiom IV we have:

13. THEOREM. The scalar field of the space is an ordered field which is dense and satisfies the Dedekind Principle.

But, it is known [7] that the only field which satisfies these conditions is the field of real numbers. (Actually, this fact can also be derived from some of the results arrived at in the later part of this section.) Hence, we get

14. COROLLARY. Up to an isomorphism, there is one and only one affine continuously ordered space, and this is just the affine space over the field of real numbers. The axiom system I, II, III', IV is consistent iff the real field is such.

Hence, we achieved the aim formulated at the beginning of this section.

c. The Axioms of Archimedes and Cantor

Various geometric studies showed the usefulness of some continuity axiom which is weaker than the Dedekind Principle. This is the so called *Axiom of Archimedes*, and we would like to consider it now.

15. DEFINITION. An ordered field K is said *to satisfy the Axiom of Archimedes* or to be an *Archimedean field* if for every $a,b \in K$ and $a > 0$ there is some natural n such that $na > b$.

Clearly, an equivalent formulation is: *for every $b \in K$ there is some n such that $n > b$ $(n = n \cdot 1)$.*

The real field is an Archimedean field. Indeed, this follows from the Dedekind Principle. Suppose, we have some $b \in K$ such that $b > n$ for every natural number n. Put then $I = \{a / a < \text{some } n\}$, $S = \{a / a > \text{every } n\}$. This defines a Dedekind cut, because S contains at least b, and the cut is defined by some number c. Then we have for every $\varepsilon > 0$:

$$c - \varepsilon < n_\epsilon < n_\epsilon + 1 < c + \varepsilon,$$

where n_ϵ is some natural number, and these relations are contradictory for $\varepsilon < 1/2$.

On the other hand, the Archimedean axiom does not imply the Dedekind Principle as we may see, for instance, from the case of the rational field Q.

The Archimedean fields are determined by

16. THEOREM. The field K is Archimedean iff it is algebraically isomorphic and similar to a subfield of the real field R.

Because the real field is Archimedean, every subfield is Archimedean as well and the condition is clearly sufficient.

Conversely, let K be an Archimedean field and $b \in K$. Then, there are some natural numbers n and m such that $b < n$ and $-b < m$, i.e., $-m < b < n$. This shows that the *rational* elements of K (i.e., elements of the form $(p \cdot 1) \cdot (q \cdot 1)^{-1}$) smaller than b define the lower class, and those greater than b define the upper class of a Dedekind cut of the rational field Q. But this is just some real number $r \in R$ and we get a map $f : K \to R$ defined by $f(b) = r$.

Now, if $a \neq b$, e.g., $a < b$, i.e., $b - a > 0$, there is a natural number $m > (b - a)^{-1}$, which means $m(b - a) > 1$. There also exists n such that $n > ma \geq n - 1$. From this relations we get $ma \geq n - (mb - ma)$, whence $mb > n$. Thus, we have $mb > n > ma$ and $b > n/m > a$ and it is easy to understand that this implies $f(a) < f(b)$ (A formal definition of the order relation between real numbers defined as Dedekind cuts in Q may be found in [7, 8].)

It follows that f is an injective map and, actually, that it is a similarity between K and $f(K)$.

As for the property of f being an algebraic isomorphism, this follows because both the operations in K and in R are defined in the same way as operations with Dedekind cuts. (For the concrete description of these operations in R see [7, 8].)

Theorem 16 is thereby proven.

17. COROLLARY. Every Archimedean field is commutative.

It follows that there also exist ordered non-Archimedean fields, e.g., the non-commutative ordered field of §3. But, such fields exist in the commutative case as well. Indeed, consider, as in §3, the field $k(R,\text{id.})$ of formal power series with real coefficients. This is an ordered commutative field which is non-Archimedean because, for instance $nt^2 < t$ for every natural number n. (The definition of $<$ for series is that of §3.)

18. COROLLARY. Every Archimedean field is separable.

Indeed, in the proof of Theorem 16, we saw that there is always some rational element between two arbitrary elements of the Archimedean field K which implies $K = \overline{Q(K)}$, where $Q(K)$ denotes the set of the rational elements of K. Q.e.d.

Since the Axiom of Archimedes does not imply the Dedekind Principle, it is interesting to get some property which, together with the Axiom of Archimedes, will imply this Principle. We start with

19. THEOREM (Cantor). Consider an ordered field K satisfying the Dedekind Principle (i.e., actually, the real field) and suppose that: a_n is an increasing sequence in K; b_n is a decreasing sequence; for every n,m we have $a_n < b_m$; for every $\varepsilon > 0$, $b_n - a_n < \varepsilon$ for all sufficiently large n. Then, there is a unique $c \in K$ such that $a_n < c < b_n$ for every natural number n.

Indeed, if we define $I = \{x/x \in K$ and $x <$ every $b_n\}$, $S = K - I$, we get a Dedekind cut on K and there is a corresponding defining element c. The relations $c > b_n$ or $c < a_n$ obviously lead to contradictions which show that c is the required element. If $d \in K$ is also such that $a_n < d < b_n$ for every n, the last hypothesis of the theorem is contradicted for $\varepsilon < |d - c|$. The only remaining possibility is $d = c$, which proves the theorem.

The condition of Theorem 19, called the *Cantor Axiom*, gives us the answer to our question, because we can prove now

20. THEOREM. A dense ordered field K satisfies the Dedekind Principle iff it satisfies both Archimedes' and Cantor's axioms.

As a matter of fact, what remained to prove is that the couple consisting of the Archimedes and Cantor axioms imply the Dedekind Principle. First, if the Axiom of Archimedes holds, we know from Theorem 16 that there is an injective isomorphism $f : K \to R$ which is also a similarity. Now, if $r \in R$, we may construct two sequences of rational numbers, satisfying the hypotheses of the Cantor Axiom and for which c of Theorem 19 is just r. Correspondingly, we obtain two similar sequences in K and, since K also satisfies the Cantor Axiom, we get a unique element $c \in K$. It is then clear that $f(c) = r$, and we see that f is also surjective. Hence K and R are isomorphic and similar, and Dedekind's Principle holds in K, q.e.d.

It should be noted that the above mentioned arguments show that the only Archimedean ordered field satisfying the Cantor Axiom is the real field R, and, implicitly, that the only ordered field satisfying the Dedekind Principle is the real field R. This fact was used in the proof of Corollary 14.

By now, we already know that the Cantor Axiom is not a consequence of the Axiom of Archimedes. The converse is also true, i.e., the Axiom of Archimedes is not a consequence of the Cantor Axiom. In fact, one can prove that the already considered non-Archimedean field $k(R,\text{id.})$ satisfies the Cantor Axiom.

Namely [6], let

$$\alpha^{(n)} = a_0^{(n)} t^{p_n} + \ldots, \quad \beta^{(m)} = b_0^{(m)} t^{q_m} + \ldots \tag{1}$$

be two sequences of formal series of $k(R,\mathrm{id.})$ which satisfy the hypotheses of the Cantor Axiom. We suppose that they are written down starting with the first non-zero coefficient.

The sequence of the exponents p_n has a lower bound since, otherwise, we have a subsequence $p_{n_k} \to -\infty$, $\alpha^{(n_k)}$ begins with a term of a higher degree in t than $\alpha^{(n_k+1)}$ and, because $\alpha^{(n_k+1)} - \alpha^{(n_k)} > 0$, we have $\alpha^{(n_k+1)} > 0$. But then, for some fixed $\beta^{(m)}$, we must have $p_{n_{k+1}} < q_m$, whence $\beta^{(m)} - \alpha^{(n_k+1)} < 0$, thereby contradicting the hypothesis that every α is smaller than every β.

Similarly, the sequence q_m has a lower bound.

Consequently, there is a smallest exponent p_n, q_m and with it we accept the convention to write every formal series of (1) starting with this smallest exponent, even if this implies having a finite number of vanishing coefficients at the beginning. That is, the sequences (1) are now

$$\alpha^{(m)} = a_0^{(m)} t^n + \ldots, \quad \beta^{(m)} = b_0^{(m)} t^n + \ldots, \tag{2}$$

where n is the above smallest exponent and, eventually, $a_0^{(m)}$, $b_0^{(m)}$ may vanish.

Further, we clearly have $b_0^{(m)} - a_0^{(m)} \geq 0$. If the strict inequality holds for every m, then take $\varepsilon = t^{n+1}$, and we get always $b_0^{(m)} - a_0^{(m)} > \varepsilon$, which is not true. Hence, a first $m = m_0$ exists such that $a^{(m_0)} = b^{(m_0)}$ and consequently $a_0^{(m)} = b_0^{(m)}$ for all $m \geq m_0$. This allows considering the real number $d_0 = a_0^{(m_0)} = b_0^{(m_0)}$. Next, we consider analogously the differences $b_1^{(m)} - a_1^{(m)}$ for $m > m_0$; they must vanish starting with some index m_1. This gives $d_1 = a_1^{(m_1)} = b_1^{(m_1)}$, and so on.

Then, the series $\alpha = d_0 t^n + d_1 t^{n+1} + \ldots$ satisfies

$$\alpha^{(m)} < \alpha < \beta^{(m)}$$

for every m, and Cantor's Axiom follows easily, q.e.d.

Now, it only remains to translate these properties into geometric language. First, we have

21. PROPOSITION. An ordered field K is Archimedean iff in the affine space over K the following property holds: let AB and CD be two collinear segments and $A < B$; let C_i $(i = 0, 1, 2, \ldots)$ be a sequence of points such that $C_0 = C$ and, for every i, $\overrightarrow{C_i C_{i+1}} \sim \overrightarrow{AB}$; then there is some index i such that $D < C_i$.

This property is meaningful because, by definition of equipollence, all C_i are on the line of the two segments. We shall define the above property as the *geometric axiom of Archimedes* and the geometries based on this axiom are called *Archimedean geometries*. The geometries in which this is not true are called *non-Archimedean geometries*.

Now, in order to prove Proposition 21, we choose in the affine space over K a frame such that the line AB is the Ox axis and $\overrightarrow{AB}$ is positively oriented. Then $A(a,0,0)$, $B(b,0,0)$, $C(c,0,0)$, $D(d,0,0)$, and, using the analytic condition for the equipollence of two vectors (1.§4), we see inductively that C_i is $(c + i(b - a), 0, 0)$. Then, the condition $D < C_i$ is equivalent to $i(b - a) > d - c$. If K is Archimedean there is some i satisfying this condition.

Conversely, if the geometric axiom of Archimedes holds, and if $a, b \in K$, then choose $A = C = (0,0,0)$, $B = (a,0,0)$ and $D = (b,0,0)$ and find i such that $C_i > D$. This means $ia > b$, and we see that K is Archimedean.

Now, from Theorem 16, Corollary 17 and the mentioned examples of non-Archimedean ordered fields we get:

22. THEOREM. The affine Archimedean geometries are just the affine geometries over fields of real numbers. In these geometries the Pappus-Pascal theorem holds. There are both non-Archimedean and non-Pappian geometries, and the Archimedean axiom is independent of the system (I, II, III').

Analogously, we can formulate a *geometric Cantor axiom* in the following way:

Let $\{A_nB_n\}$ $(n = 0,1,2...)$ be a strictly including sequence of segments (like in Corollary 7) and suppose that for every segment PQ of the same line there is some n_0 such that $n > n_0$ and $\overrightarrow{PX_n} \sim \overrightarrow{A_nB_n}$ or $\overrightarrow{PX_n} \sim \overrightarrow{B_nA_n}$ (where we choose that of the equipollences for which X_n and Q are on the same half-line with the origin P) imply $P - X_n - Q$. Then there is a unique point M such that, for every n, $A_n - M - B_n$.

The reader can prove without difficulty, and employing the same method as in the case of Proposition 21, that the geometric Cantor Axiom is equivalent to the fact that the corresponding scalar field satisfies the Cantor Axiom of Theorem 19.

If we denote by IV' 1 the geometric axiom of Archimedes and by IV' 2 the geometric Cantor Axiom, then Theorem 20 yields

23. THEOREM. The axiom systems (I, II, III', IV) and (I, II, III', IV') are equivalent.

This provides us with a second answer to the problem of characterizing the real affine space.

Finally, the example of the field $k(R,\text{id.})$ indicates that we have:

24. THEOREM. The Axiom of Archimedes is independent of the axioms I, II, III', and IV' 2.

Additional problems related to the continuity axioms will be considered in the next chapter.

PROBLEMS

1. Let A,B,C,D be four collinear points in a geometrically ordered space. Prove that the conditions $A - D - C$ and $B - D - C$ imply $\overline{A - D - B}$.

2. Let A,B,C,D be four collinear points in a geometrically ordered space. Then, if $A - B - C$ and $B - C - D$ it follows that $A - B - D$ and $A - C - D$.

3. Let A,B,C,D be four collinear points and suppose $A - B - C$ and $A - C - D$. Show that this implies $B - C - D$ and $A - B - D$.

4. Show that whenever we have n collinear points it is possible to choose an order of these points such that if the points in this order are $A_1, A_2, \ldots, A_n$ then $A_i - A_j - A_k$ iff either $i < j < k$ or $i > j > k$.

5. Prove, without Theorem 5 but using the results of the foregoing problems, that given a point O on a line d, the other points of d may be uniquely divided into two classes such that two points belong to different classes iff O lies between these two points.

6. Use the result of Problem 5 to give a new proof of Theorem 1.8 of this chapter.

7. Give a detailed proof of the case 3° of Proposition 1.15.

8. Let u,v,w be three coplanar half-lines with the same origin O and belonging to different half-lines. Then, show that either any two of them are in different half-planes with respect to the line containing the third or there is some line not containing O and intersecting u, v and w.

9. Give a detailed proof of the fact that four half-lines of a pencil may be uniquely grouped into two separating pairs.

10. If s,u,v,w are half-lines of the same pencil and (s,u) separates (v,w), then show that (v,w) separates (s,u) as well. In this case we say that one has the *cyclic order* (s,v,u,w).

11. If s,u,v,w are half-lines of the same pencil, then one and only one of the following cyclic orders exists: (s,u,v,w), (s,v,u,w), (s,u,w,v). If one has n half-lines of a pencil, it is possible to denote them in some order $u_1,\ldots,u_n$, such that (u_i,u_j,u_k,u_l) if $i<j<k<l$.

12. Formulate and prove an orientability theorem, like Theorem 1.5, for $F(0) - \{u\}$, where $F(0)$ is a pencil of half-lines and $u \in F(0)$.

13. Let $A_1 \ldots A_n$ be a convex polygon and D,E two of its points, which do not belong to the same side. Show that every interior point of the segment DE is in the interior of the polygon and that the corresponding polygonal plate is decomposed into a convex sum of two plates.

14. Give a detailed proof of Theorem 2.9.

15. Prove that in any geometrically ordered plane there is a simple and non-convex polygon.

16. Show that if a polygonal sub-division of an n-gon has $n+2$ vertices, then it must have $n+2$ faces and $2n+3$ edges.

17. Show that Fig. 51 actually defines a polyhedron and show that this polyhedron is not simply connected.

18. Construct polyhedra for which the number of vertices (v), the number of edges (m) and the number of faces (f) satisfy the condition $v - m + f = -2,-4,-6,\ldots,-2p,\ldots$.

19. Consider a geometrically ordered space satisfying the Hilbert axioms III'. Show that in this case the ordered space topology is characterized by the fact that the interiors of the tetrahedra give a basis of open neighborhoods of the space. Which induced bases do we have in a plane and on a line? Show that the mentioned topology is Hausdorff.

20. Give a detailed proof of Corollary 3.4.

21. Prove that in an ordered affine space the diagonals of any parallelogram have an intersection point.

22. Let α be a geometrically ordered plane, d_1 and d_2 two lines in α and M a point of α which belongs neither to d_1 nor to d_2. Does the projection of d_1 to d_2 from M preserve the betweenness relation?

23. Let (A,B) and (C,D) be harmonic collinear pairs in an ordered affine space (see problem 29 of Chapter 1). Prove, using coordinates, that if $A - C - B$, then $\overline{A - D - B}$.

24. A subset M of an ordered space is called *convex* iff $P,Q \in M$ and $P - R - Q$ imply $R \in M$. Show that for any subset U of the space there is a smallest convex subset M containing U. This is called the *convex envelope* of U.

25. Let M be a convex set and O be an arbitrary point of the space. Show that the union of the segments OX for $X \in M$ is again a convex set.

26. A subset M of an ordered affine space is said to be *bounded* if there is some tetrahedral body T such that every point of M is either interior or on the boundary of T. Show that every finite subset is bounded.

27. Show that M is a bounded set iff there is a parallelepiped P such that every point of M is either interior or on the boundary of P.

28. Let α be the intuitive Euclidean plane and γ the interior of a circle of α. Define a plane having as points the points of γ and as lines open circle arcs in γ with the ends in two diametrically opposite points on the boundary of γ. Show that this is an ordered affine plane.

29. Compute arbitrary determinants of order 2 and 3 over the quaternion field. (See Problem 28 of Chapter 1).

30. Give a detailed proof of Theorem 3.10.

31. Prove that any affine ordered plane in which the Pappus Theorem holds is an Arguesian plane.

32. Let $(M,<)$ be a totally ordered dense set and S a subset of M. Show that S is the upper class of a Dedekind cut in M iff the

following properties hold: 1°: S and $M-S$ are nonvoid sets; 2°: $x \in S$ and $x' > x$ implies $x' \in S$; 3°: $x \in S$ implies the existence of $x' \in S$ such that $x' \in S$. (It is accepted here that the defining element, if it exists, is the greatest in the lower class.)

33. Prove that any infinite bounded subset of points of a continuous ordered affine space (see Problem 26 above) has a limit point.

34. Prove Corollary 4.10. Derive from it that under its hypotheses and if the space is also affine it is homeomorphic to R^3.

35. Let A,B,C be three noncollinear points of a continuous ordered plane α. Show that there are in α open neighborhoods of these points: U_A, U_B, U_C such that if $A' \in U_A$, $B' \in U_B$, $C' \in U_C$ then A', B', C' are noncollinear points. A similar property holds for four noncoplanar points in the space.

36. Prove the equivalence between the geometric Cantor Axiom and the Cantor Axiom in the scalar field.

37. Formulate and prove, in a ordered affine space satisfying the Cantor Axiom, a proposition similar to the Cantor Axiom in which the strictly including segments are replaced by strictly including parallelepipeds with parallel edges.

38. Give new examples of ordered affine Archimedean geometries in which Cantor's axiom does not hold.

CHAPTER 3
EUCLIDEAN SPACES

§1. THE CONGRUENCE AXIOMS AND THEIR RELATIONS WITH THE INCIDENCE AND ORDER AXIOMS

a. A Preliminary Discussion of Congruence

If you recall, our basic aim is to obtain at the final stage all the properties encountered in physical space. But among these properties we also have the possibility of considering equality of geometric figures, and motions in space, and these subjects have no adequate analogues in ordered affine spaces even if these are continuous.

To overcome this difficulty, we shall introduce a new fundamental binary relation, on the set of nonoriented segments, which is called the *congruence* of segments. This relation has to be characterized by convenient axioms which, also, will relate congruence to the already known geometrical relations. Thus, we get a fifth group of axioms called *congruence axioms*, which we shall introduce gradually in the sequel.

We begin with the following first axiom

V 1. The congruence of segments is an equivalence relation.

That is, if we denote by $AB \equiv CD$ the congruence of two segments, then $AB \equiv AB$ (reflexivity), $AB \equiv CD$ implies $CD \equiv AB$ (symmetry) and $AB \equiv CD$, $CD \equiv EF$ imply $AB \equiv EF$ (transitivity).

As always, "equality" allows for defining "motions," which, in our case, will be done as follows

1. DEFINITION. A bijective point transformation of the space onto itself, or of a line onto a line, or of a plane onto a plane, is called a *congruent transformation* (shortly, a *congruence*) if it sends every segment to a congruent segment.

These congruences are just the motions referred to and we can prove

2. PROPOSITION. The set of space congruences forms a transformation group. The same holds for the congruences of a line or a plane onto themselves.

In fact, it follows from Definition 1 that the inverse of a congruence and the composition of two congruences are again congruences.

This group will be denoted by E (or $E(\alpha)$,$E(d)$) and, in this book, we shall call it the *Euclidean group*.

Another useful notion is introduced by

3. DEFINITION. Let $(A_1,\ldots,A_n)$, $(B_1,\ldots,B_n)$ $(n \geq 2)$ be two ordered groups of n points. They are called *congruent*, which is denoted by $(A_1,\ldots,A_n) \equiv (B_1,\ldots,B_n)$, if $A_iA_j \equiv B_iB_j$ for every couple of indices i,j.

Clearly, a congruence sends every group of n points to a congruent group.

So far, no relation has been established between congruence and the already existing structure of the space so that, if we refer to pairs of arbitrary elements instead of segments, Axiom V 1 is still meaningful. Hence, the notion of a congruence will be of interest to us only after we establish some compatibility relations with the structures of the foregoing chapters.

First, we introduce the following axioms which relate congruence to the lines-planes structure of the space:

V 2. If $(A,B,C) \equiv (A',B',C')$ and A,B,C are collinear points, then A',B',C' are collinear as well.

V 3. If $(A,B,C,D) \equiv (A',B',C',D')$ and A,B,C,D are coplanar points, then A',B',C',D' are coplanar as well.

Now, we can examine a geometry based on the Axioms I and V. For instance, we have

4. THEOREM. A congruence of the space sends lines onto lines and planes onto planes.

Indeed, if d is a line, we take on it two points A,B and consider their images A',B'. By V 2, every point $C \in d$ is sent to a point of the line A',B' and conversely. The assertion for planes can be obtained in a similar manner.

However, such a geometry is too poor.

Suppose that we also add the parallel axiom II. Then we get from Theorem 4

5. COROLLARY. In an affine space with congruence, the Euclidean group is a subgroup of the affine group.

This geometry is not too rich either.

As a consequence, we are led to add the order axioms too. Moreover, in this section we shall study first the geometry based on incidence, order and congruence axioms. Then, in the following section, we shall add to it the parallel axiom.

b. Elementary Properties of Congruence of Segments

Our convention here is *to use the order axioms III'*.

We shall add some compatibility axioms between congruence and order, which lead to interesting geometrical consequences. Namely, we consider first

V 4. Let (A,B,C) and (A',B',C') be two triples of points, such that: $A - B - C$, $A' - B' - C'$, $AB \equiv A'B'$, $BC \equiv B'C'$. Then, we also have $AC \equiv A'C'$.

V 5. Consider some segment AB, a point A' and a half-line s with the origin A'. Then, there is on s a unique point B' such that $AB \equiv A'B'$.

Both these axioms have a simple intuitive significance. V 4 is, somehow, the converse of axiom V 2. Axiom V 5 is a strong axiom because it requires the existence of "sufficiently many" points. It is also worth noting that Axiom V 5 allows for giving a weaker form for V 1, namely:

V $\bar{1}$. If $A'B' \equiv AB$ and $A''B'' \equiv AB$, then $A'B' \equiv A''B''$.

In fact, V 5 allows us to construct for every segment AB some segment $A'B'$ such that $AB \equiv A'B'$. If we write down this relation twice and use V $\bar{1}$, we get $AB \equiv AB$. Next, from $A'B' \equiv A'B'$ and $AB \equiv A'B'$, we get $A'B' \equiv AB$ by V $\bar{1}$. And finally, the symmetry of the congruence relation, together with V $\bar{1}$, obviously implies the transitivity of $\equiv$.

For the moment, we want to study the consequences of I, III', and V 1-5, and we begin with some results which complement Axiom V 4.

6. PROPOSITION. Let (A,B,C) and (A',B',C') be two triples of collinear points. Then, if $A - B - C$, $\overline{B' - A' - C'}$, $AB \equiv A'B'$ and $AC \equiv A'C'$, we also have $A' - B' - C'$ and $BC \equiv B'C'$.

In fact, according to Axiom V 5, there is a unique point C'' such that $A' - B' - C''$ and $BC \equiv B'C''$. But then, V 4 yields $AC \equiv A'C''$ and C',C''

are on the same side of A'. Hence, because of the uniqueness assertion in Axiom V 5, $C'' = C'$, and Proposition 6 is proven.

7. PROPOSITION. If $(A,B,C) \equiv (A',B',C')$ and $A - B - C$ then $A' - B' - C'$.

Indeed, the relation $A - B - C$ implies that these three points are collinear whence, by V 2, A',B',C' are collinear as well. Further, the desired result follows by a reductio ad absurdum and the use of Proposition 6.

8. COROLLARY. The congruent transformations preserve the order of points on a line. They are sending half-lines, half-planes and half-spaces to half-lines, half-planes and half-spaces respectively and either preserve or invert orientations.

This corollary allows for classifying congruences into *direct congruences* or *motions*, which preserve the orientation and *indirect congruences* which invert the orientation.

Another consequence of Proposition 6 and axiom V 4 is:

9. PROPOSITION. Let A,B,C be three collinear points and A',B' two points such that $AB \equiv A'B'$. Then there is a unique point C' such that $(A,B,C) \equiv (A',B',C')$.

Indeed, according to V 2, we must look for such a C' on the line $A'B'$. If $A - B - C$, C' is defined, in view of V 5, by the conditions $A' - B' - C'$ and $BC \equiv B'C'$ and, because of V 4, it provides us with the answer to our problem. Similar is true if $C - A - B$. Finally, if $A - C - B$, then we define C' by the conditions that it is on the same half-line with origin A' as B' and $AC \equiv A'C'$, and use Proposition 6.

It is of interest to consider now:

10. DEFINITION. The *congruence middle* of a segment AB is a point M of the line AB such that $AM \equiv MB$.

We shall discuss later the existence of such a point. At this moment, let us prove

11. PROPOSITION. The congruence middle M of a segment AB (if any) is unique and one has $A - M - B$.

First, if $\overline{A - M - B}$, then we have, on some half-line with the origin M, $MA \equiv MB$ in contradiction to Axiom V 5. Thus, $A - M - B$ follows necessarily.

Next, if M' is also a congruence middle of AB and if, for instance, $A - M - M'$, then, using the order $A < B$, we see that $B - M' - M$, which contradicts Corollary 8, q.e.d.

In order to make some other important considerations, let us note by Segm the set of all segments of space, and by [Segm] the corresponding quotient set of classes of congruence segments. Then:

12. PROPOSITION. [Segm] has a canonical structure of an abelian additive semigroup and it has also a canonical order.

In fact, denoting by $[AB]$ the congruence class of the segment AB, we may define $[AB] + [CD] = [MN]$, where MN is obtained in the following manner: take an arbitrary half-line s with the origin M and, by using V 5, define on s the point P such that $AB \equiv MP$; next, on s, and on the opposite side of M with respect to P, take N such that $CD \equiv PN$. The independence of the result on the choice of M, s, AB, CD (in the same congruence classes, of course) is a trivial consequence of V 5 and V 4. Further, associativity is straightforward, and commutativity follows by using V 4 once again. Hence, we obtain the desired semigroup structure.

As for order, we define $[AB] < [CD]$ if, when we construct on the half-line from C to D a point M such that $AB \equiv CM$, this point M satisfies the relation $C - M - D$. The reader will prove by himself the independence of this definition of the choice of the representative segments of the given congruence classes and the fact that < is in fact an order relation.

Let us also note that, if $[AB] < [CD]$, we may define $[CD] - [AB] = [MD]$, where M is the point which was constructed before.

Usually, the operations +, - and the relation < are written for segments (i.e., omitting brackets) and one speaks of a sum, difference or order of segments. This is an abuse of notation and language but, obviously, it causes no confusion.

c. The Euclidean Group of a Line

In the sequel, we fix some line d and consider exclusively its segments. Moreover, we also fix some orientation of d. Then, it is meaningful to define that two oriented segments are *equal* if they are congruent and have the same orientation.

The equality relation will be denoted by = and the classes of equal oriented segments of d will be called *congruence vectors* on d.

Note that Axiom V 5 assures for the congruence vectors the validity of Property I.3.6 of "parallelism" vectors, for points of d. It follows that we may use the definition of 1, §3 for the sum of vectors and, thereby, organize the set of congruence vectors of d as an additive abelian group denoted by $V_c(d)$. The reader is invited to provide a full proof of this fact.

Here we shall use congruence vectors to study the structure of the Euclidean group $E(d)$ of the line d. Since all the vectors of this section are congruence vectors, we shall omit hereafter the word "congruence" and say simply "vectors."

We begin with the following result:

13. THEOREM. Let A,B be two points of a line d and A',B' two points of some line d' (which, possibly coincides with d) such that $AB \equiv A'B'$. Then, there is one and only one congruent transformation of d onto d', which sends A to A' and B to B'.

First, if such a transformation exists and is $f: d \to d'$, we have necessarily $(A,B,C) \equiv (A',B',C')$, where $C' = f(C)$, and f is unique by Proposition 9.

As for the existence, define $f: d \to d'$ by $f(C) = C'$ determined by Proposition 9 and show that this f is a congruent transformation.

Indeed, take two points $C,D \in d$ and assume, for instance, that $A < B < C < D$ and $A' < B'$.

Then we have, by Corollary 8, $A' < B' < C'$ and $A' < B' < D'$, i.e., C' and D' are on the same side of B'. This fact, together with the relations $BC \equiv B'C'$, $BD \equiv B'D'$, which hold in view of the definition of f, yield, by Proposition 6, $CD \equiv C'D'$.

In a similar manner we prove this relation for the other possible positions of the points C,D. Hence, f is a congruence, q.e.d.

14. DEFINITION. Let A,B be two points of an oriented line d. The *congruent translation* defined by the vector $\overrightarrow{AB}$ or by the pair (A,B) is the transformation $T^c_{\overrightarrow{AB}}: d \to d$ defined by: $T^c_{\overrightarrow{AB}}(P) = P'$ if $\overrightarrow{PP'} = \overrightarrow{AB}$. The *symmetry* defined by the vector $\overrightarrow{AB}$ or by the pair (A,B) is the transformation $S_{\overrightarrow{AB}}: d \to d$ defined by: $S_{\overrightarrow{AB}}(P) = P'$ if $\overrightarrow{AP} = \overrightarrow{P'B}$.

Since the only translations of this section are the congruent translations, we omit hereafter the word congruent.

We shall note a few simple consequences.

If $P' = T^c_{\overrightarrow{AB}}(P)$, then $T^c_{\overrightarrow{PP'}} = T^c_{\overrightarrow{AB}}$. Indeed, this follows because equality of oriented segments is an equivalence relation.

Similarly, if $P' = S_{\overrightarrow{AB}}(P)$, then $S_{\overrightarrow{PP'}} = S_{\overrightarrow{AB}}$. Indeed, from $\overrightarrow{AP} = \overrightarrow{P'B}$ and $\overrightarrow{PQ} = \overrightarrow{Q'P'}$ we get $\overrightarrow{AQ} = \overrightarrow{AP} + \overrightarrow{PQ} = \overrightarrow{Q'P'} + \overrightarrow{P'B} = \overrightarrow{Q'B}$, and something similar happens if we change the role of $\overrightarrow{AB}$ and $\overrightarrow{PP'}$.

Hence, we can say that a translation and a symmetry are defined by an arbitrary pair of corresponding points.

Now, if the segment AB has the congruence middle M and if $P' = S_{\overrightarrow{AB}}(P)$ then PP' also has the middle M. Indeed, we have $\overrightarrow{MA} + \overrightarrow{AP} =$

$\overrightarrow{BM} + \overrightarrow{P'B}$, i.e., $\overrightarrow{MP} = \overrightarrow{P'M}$. It follows that the symmetry is characterized by the relation $\overrightarrow{MP} = \overrightarrow{P'M}$ and we can speak of a *symmetry with respect to the point M.*

If we consider $T_{\overrightarrow{AB}}$ and $T_{\overrightarrow{CD}}$, we get, for corresponding points,

$$\overrightarrow{PP'} = \overrightarrow{AB}, \quad \overrightarrow{P'P''} = \overrightarrow{CD}, \quad \overrightarrow{PP''} = \overrightarrow{PP'} + \overrightarrow{P'P''} = \overrightarrow{AB} + \overrightarrow{CD},$$

which proves that the composition of two translations is again a translation and, moreover, that the translations form a group which is isomorphic to $V_c(d)$.

Finally, we note that the product of two symmetries is a translation, because, if $P' = S_{\overrightarrow{AB}}(P)$ and $P'' = S_{\overrightarrow{CD}}(P')$, we have $\overrightarrow{AP} = \overrightarrow{P'B}$, $\overrightarrow{CP'} = \overrightarrow{P''D}$ from which it follows $\overrightarrow{PP''} = \overrightarrow{PA} + \overrightarrow{AB} + \overrightarrow{BP'} + \overrightarrow{P'C} + \overrightarrow{CD} + \overrightarrow{DP''} = \overrightarrow{AB} + \overrightarrow{CD} + 2\overrightarrow{BC} =$ const. Particularly, $S^2_{\overrightarrow{AB}} = T_{\overrightarrow{AB} + \overrightarrow{AB} + 2\overrightarrow{BA}} =$ identity, which means, by definition, that the symmetries are *involutive transformations.*

We suggest to the reader to show, in an analogous manner, that the product of a symmetry and a translation is a symmetry.

Now, we may obtain the main result

15. THEOREM. The translations are direct congruent transformations of the line and the symmetries are indirect congruent transformations. Every element of $E(d)$ is either a translation or a symmetry according to whether it is a direct or an indirect congruence.

In fact, take $T_{\overrightarrow{AB}}(P_i) = P_i'$ $(i = 1,2)$. Then $\overrightarrow{P_1'P_2'} = \overrightarrow{P_1'P_1} + \overrightarrow{P_1P_2'} + \overrightarrow{P_2P_2'} = \overrightarrow{BA} + \overrightarrow{P_1P_2} + \overrightarrow{AB} = \overrightarrow{P_1P_2}$, which obviously shows that we have a direct congruence.

Similarly, if $S_{\overrightarrow{AB}}(P_i) = P_i'$ $(i = 1,2)$, we have

$$\overrightarrow{P_1'P_2'} = \overrightarrow{P_1'P_2} + \overrightarrow{P_2P_1} + \overrightarrow{P_1P_2'} = \overrightarrow{P_1'A} + \overrightarrow{AB} + \overrightarrow{BP_2} + \overrightarrow{P_2P_1} + \overrightarrow{P_1B} + \overrightarrow{BA} + \overrightarrow{AP_2'} = \overrightarrow{P_2P_1}$$

(after reductions), i.e., we have an indirect congruence.

As for the second part of the theorem, if $M \in E(d)$ and it is a direct congruence, then, for corresponding points A,A' and B,B', we have $\overrightarrow{AB} = \overrightarrow{A'B'}$, whence $\overrightarrow{AA'} = \overrightarrow{AB} + \overrightarrow{BA'} = \overrightarrow{BA'} + \overrightarrow{A'B'} = \overrightarrow{BB'}$, whence $M = T_{\overrightarrow{AA'}}$. If M is an indirect congruence then $\overrightarrow{AB} = \overrightarrow{B'A'}$ which, by the very definition of a symmetry, means $M = S_{\overrightarrow{AA'}}$.

Theorem 15 is thereby completely proven.

d. Plane Congruence Properties

So far we have been getting congruence properties on a line. In order to get plane and spatial properties, we need some more assumptions and we shall introduce the following axioms:

V 6. Let A,B,C be three noncollinear points and consider some segment $A'B' \equiv AB$ and a half-plane s with the line $A'B'$ as origin. Then, there is a unique point C' in s such that $(A,B,C) \equiv (A',B',C')$.

V 7. Let A,B,C be three noncollinear points, P be a point of the line AB and A',B',C',P' be four points such that $(A,B,C) \equiv (A',B',C')$ and $(A,B,P) \equiv (A',B',P')$. Then it follows that $CP \equiv C'P'$.

Now, some of the previous congruence axioms become logically dependent. Namely:

16. PROPOSITION. If the axiom groups I and III' are assumed, Axioms V 2 and V 3 are consequences of the other congruence axioms introduced.

We prove this only for V 2 and avoid the rather lengthy proof of V 3 [10].

Let us consider the points $A - B - C$ and $(A',B',C') \equiv (A,B,C)$ and suppose that A',B',C' are not collinear (Fig. 63). Then, choose a half-plane with the origin line AB and take in it the unique point

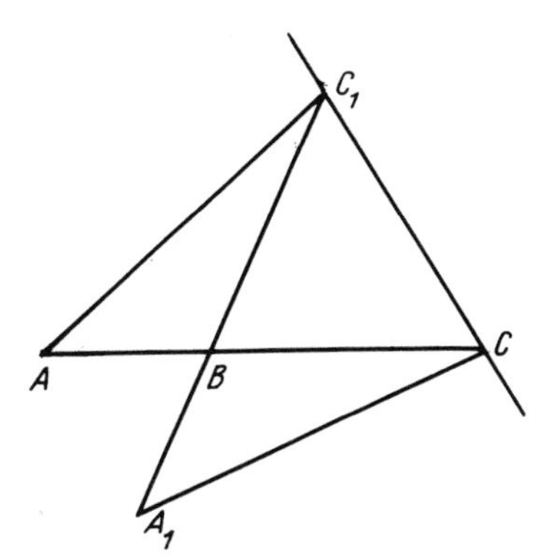

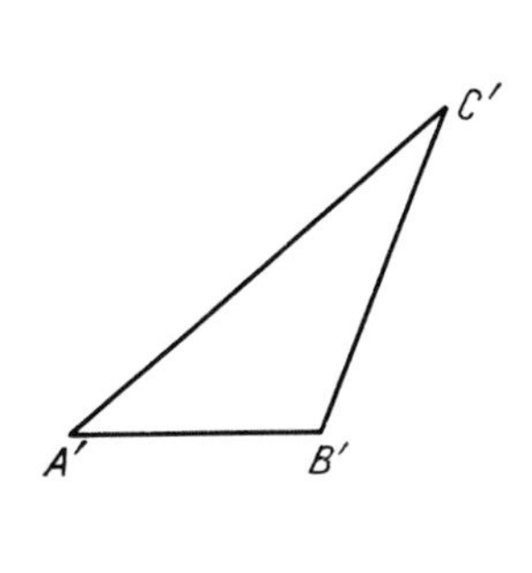

Fig. 63

C_1 such that $(A,B,C_1) \equiv (A',B',C')$. Next, take A_1 on BC_1 such that $(A,B,C) \equiv (A_1,B,C_1)$.

Then, we clearly have $(A,B,A_1) \equiv (A_1,B,A)$, which, together with the condition defining A_1 and with Axiom V 7, imply $A_1C \equiv AC_1$.

It follows $(A,C,C_1) \equiv (A_1,C,C_1)$ and, because A and A_1 are in the same half-plane with respect to CC_1, we have a contradiction with Axiom V 6.

The assertion of Proposition 16 follows thereby.

Presently, we can prove some other result:

17. PROPOSITION. Every segment has a congruence middle.

Let AB be an arbitrary segment and C a point which is noncollinear with A,B. In the plane ABC, and on the side of the line AB not containing C, take C' defined (in view of V 6) by $(A,B,C) \equiv (B,A,C')$. Denote by P the intersection point of AB and CC' (Fig. 64). Then, take on the line AB the point P' defined by the condition $(A,B,P) \equiv (B,A,P')$. From the relations $(A,B,C) \equiv (B,A,C')$ and $(A,B,P) \equiv (B,A,P')$ we have, by V 7, $CP \equiv C'P'$. From the relations $(A,B,C) \equiv (B,A,C')$ and $(A,B,P') \equiv (B,A,P)$ (which can be easily established), we get $CP' \equiv C'P$. Hence, $(C,P,C') \equiv (C',P',C)$ and C,P',C' are collinear. It follows that

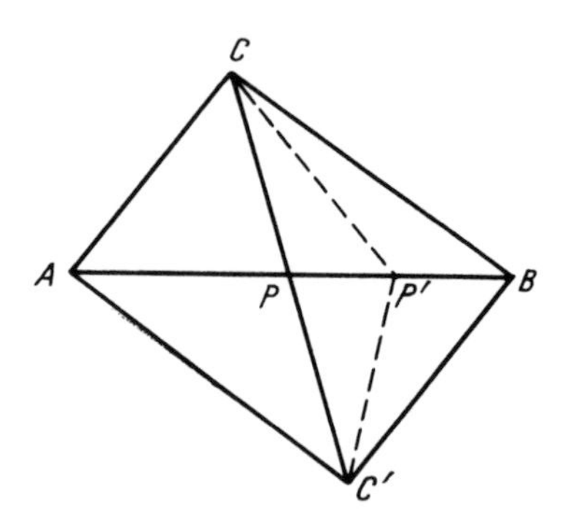

Fig. 64

P' coincides with P, and, clearly P is the congruence middle of the segment AB, q.e.d.

An important plane congruence property is

18. THEOREM. Let α and α' be two planes (possibly $\alpha = \alpha'$) and $(A,B,C) \equiv (A',B',C')$ two noncollinear congruent triples of α and α' respectively. Then, for any $P \in \alpha$, there is some point $P' \in \alpha'$ which is uniquely defined by the condition $(A,B,C,P) \equiv (A',B',C',P')$. If Q,Q' is another similar pair of corresponding points, then $PQ \equiv P'Q'$.

Indeed, if P belongs to a side of the triangle ABC, P' is defined by applying Proposition 9 to this side and its corresponding side.

If this is not the case, then, for instance, AP intersects BC at a point D (Fig. 65) and we get P' by twice applying Proposition 9, first for D on BC which defines a corresponding point D' on $B'C'$, and second to find P' on $A'D'$. The proof of the fact that P' is actually the required point follows from repeated applications of Axiom V 7. The uniqueness of P' follows with a reductio ad absurdum and from the use of Axiom V 6.

The second part of the theorem follows as well from V 7 and it is left as an exercise to the reader.

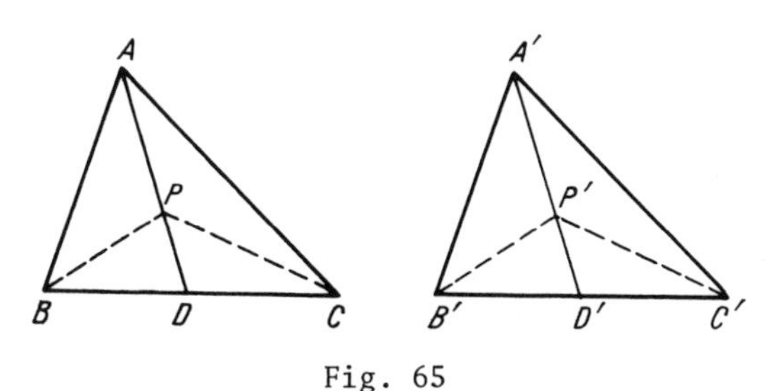

Fig. 65

We can sketch now the structure of the group $E(\alpha)$ of congruent transformations of a plane α.

First, from Theorem 18 we have:

19. COROLLARY. Let α and α' be two planes and $(A,B,C) \equiv (A',B',C')$ two noncollinear congruent triples in these planes, respectively. Then, there is one and only one congruent transformation between the two planes which sends A to A', B to B' and C to C'.

The proof is similar to that of Theorem 13.

20. DEFINITION. Let $a \subseteq \alpha$ be a line and A,B two points of a. Define for every point $P \in \alpha$ the point $P' \in \alpha$ such that: 1° $P' = P$ if $P \in a$; 2° $(A,B,P') \equiv (A,B,P)$ if $P \bar{\in} a$. Then, the correspondence $P \leftrightarrow P'$ is called the *symmetry of* α *with respect to the axis* a.

This is a well defined notion, because, if A and B are given, P' is well-defined by Axiom V 6, and if we take instead of A,B the points $A',B' \in a$, Axiom V 7 implies the fact that we obtain the same point P'.

21. THEOREM. The symmetries in α are involutive transformations. A symmetry is uniquely defined by an arbitrary pair of corresponding distinct points.

The first assertion is clear: the composition of a symmetry with itself is the identity mapping.

For the second assertion, we have to show that, if we have two arbitrary points P and P' of a plane α, then there is a unique line in α such that P and P' are points symmetric with respect to this line.

Indeed, let us consider some point $Q \in \alpha$ and define $Q' \in \alpha$ by the condition $(Q,P,P') \equiv (Q',P',P)$ (Fig. 66). Then, let A be the middle of PP', B the middle of QQ' (or $B = Q = Q'$), and a be the line AB. From $(P,Q,Q') \equiv (P',Q',Q)$ and $QB \equiv Q'B$ we deduce by V 7, $PB \equiv P'B$ which, together with $PA \equiv P'A$, shows that a is a symmetry axis for the points P,P'.

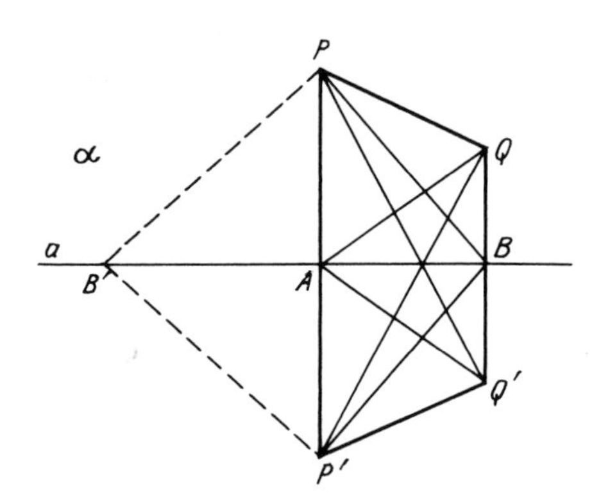

Fig. 66

Further, every symmetry axis a' of P and P' must obviously contain the middles of the segments PP' and QQ' above, whence $a' = a$, and the theorem is proven.

22. REMARK. The symmetry with respect to the axis PP' preserves the line a (the notation is the same as in the last theorem).

Indeed, if B' is the point symmetric to B with respect to the axis PP' (Fig. 66), then $PB' \equiv P'B'$ and the line AB' is a symmetry axis for the pair P,P'. This implies that the line AB' coincides with a and $B' \in a$. Because B is an arbitrary point of a, our assertion follows.

23. THEOREM. The symmetries are indirect congruences of the plane. Every indirect congruence of a plane, which has at least two fixed points is a symmetry with respect to the line joining these points.

Using the notation of Definition 20, let P and P' be two points symmetric with respect to the line a and $A,B \in a$. According to Corollary 19, there is a unique congruence which sends A to A, B to B and P to P'. Clearly, if Q and Q' are also symmetric, they correspond to each other by this congruence and conversely. This shows that the symmetry with respect to a is just the mentioned congruence, and this is an indirect congruence because the corresponding triangles ABP and ABP' have different orientations (Theorem 2,§1.17). This proves the first part of Theorem 23.

As for the second part, in view of Axiom V 6, the existence of the two fixed points implies that the transformation must be either the mentioned symmetry or the identity map. But the identity is a direct transformation, which is not the case here, q.e.d.

As a result, the structure of the group $E(\alpha)$ is given as follows:

24. THEOREM. Every direct congruence of the plane is the product of two symmetries. Every indirect congruence is either a symmetry or the product of three symmetries.

Indeed, let T be a direct congruence in α. If $T = \text{id.}$ then it may be considered as the square of an arbitrary symmetry. If $T \neq \text{id.}$, consider such a point A that its image $T(A) = B \neq A$, and introduce also the point $C = T^{-1}(A)$.

By Theorem 21, we have a symmetry axis d of the points C and A and a symmetry axis d' of the points C and B. Moreover, d' contains A (by the construction in the proof of Theorem 21), and, if $B = C$, d' is just the line AB.

If we denote by S_d and $S_{d'}$ the symmetries with axes d and d' respectively, we see that $S_{d'}S_d(C) = A$, $S_{d'}S_d(A) = B$. Hence, $T^{-1}S_{d'}S_d(C) = C$ and $T^{-1}S_{d'}S_d(A) = A$. This shows that $T^{-1}S_{d'}S_d$ is a direct congruence with two fixed points and, as we saw in the proof of Theorem 23, it must be the identical transformation. It follows $T = S_{d'} S_d$ as requested.

Now, suppose T is an indirect congruence which is not a symmetry, and let S be an arbitrary symmetry. Then ST is a direct congruence and we have $ST = S'S''$ where S' and S'' are two symmetries. It follows that $T = S^{-1}S'S''$ which completes the proof of Theorem 24.

We would like to note in passing that we may also introduce the symmetries with respect to a point, which are direct congruences, by means of the classical definition. It is also possible to define translations and rotations (as we shall see later) and to use them in order to obtain other structure theorems for the group E.

Similar developments may be considered in the three-dimensional case (e.g., [10]). Here, one has an analogous definition of the symmetries with respect to planes and it follows that the elements of of E are products of such symmetries.

e. Miscellaneous Congruence Properties

In the remaining part of this section, we would like to discuss various other geometrical consequences of Axioms I, III' and V.

Consider first

25. DEFINITION. Let a,b be two coplanar lines. If a is a fixed line of the symmetry with the axis b, we say that a is *perpendicular (orthogonal)* to b and we denote $a \perp b$.

26. PROPOSITION. If $a \perp b$, a and b are concurrent lines and also $b \perp a$.

Both assertions follow from the definition of symmetry.

Now, the usual theory of orthogonal lines can be obtained, and the reader is called upon to prove, as an exercise, facts such as: through any point there is one and only one perpendicular to any line which does not contain the point; the congruent transformations send orthogonal lines to orthogonal lines; if in each of the planes α, α' we have a pair of orthogonal lines (d_1, d_2) and (d_1', d_2') respectively, there is some congruent transformation $f : \alpha \to \alpha'$ which sends d_1 to d_1' and d_2 to d_2'.

Analogously, we can define the perpendicularity of a line l and a plane α ($l \perp \alpha$) through the condition that l be a fixed line of the symmetry with respect to α. This allows for establishing the classical theorems on perpendicular lines and planes.

The perpendicularity relation may be used to define congruence translations in a plane and in space. Namely, in a plane α, the translation defined by the pair of points A, A' is the transformation which associates with every point $P \in AA'$ the point $P' = T_{\overrightarrow{AA'}}(P)$ such that $\overrightarrow{AA'} = \overrightarrow{PP'}$ (which is meaningful because we are on a line), and to every point $P \bar{\in} AA'$ the point P' defined as follows: take the intersection Q of the orthogonal line from P to AA', take $Q' = T_{\overrightarrow{AA'}}(Q)$ as above, take the line d of α which passes through Q' and is perpendicular to AA', finally, take P' on d in the same half plane of AA' as P and such that $Q'P' \equiv QP$. A similar definition is obviously available in the three-dimensional case. The translations are direct congruences.

Another interesting subject is the classical theory of congruent angles and triangles.

27. LEMMA. If $\widehat{hOk}$ is an angle of the plane α, there is a unique symmetry of the plane α which sends the half-line h onto k and conversely.

Indeed, take $A \in h$ and $B \in k$ such that $OA \equiv OB$ and let M be the congruence middle of the segment AB. It easily follows that OM is the axis of the required symmetry. If there is one more axis with the same property, this axis must contain O (which is a fixed point of the symmetry) and must be in the interior of $\widehat{hOk}$. Hence, by Proposition 2.2.15, it must intersect AB in a point N. By the definition of a symmetry $NA \equiv NB$. It follows $N = M$ (uniqueness of the middle). Hence the two symmetry axes coincide, q.e.d.

28. DEFINITION. The symmetry axis of Lemma 27 is called the *interior bisector of the angle* $\widehat{hOk}$ and the line which contains O and is orthogonal to the interior bisector in the plane of the angle is called the *exterior bisector of* $\widehat{hOk}$.

Lemma 27 shows that every angle has a unique interior and a unique exterior bisector.

29. DEFINITION. Two angles $\widehat{hOk}$ and $\widehat{h'O'k'}$ are called *congruent* if there is a congruent transformation between their planes which sends O to O', h onto h' and k onto k'.

Note that, if we have some congruence which sends O to O', h to k' and k to h', then the composition of this transformation with the symmetry with respect to the interior bisector of the angle $\widehat{h'O'k'}$ gives a congruence like in Definition 29. Hence, *the congruence relation is in fact a relation for nonoriented angles*.

As an example of congruent angles, we have the two angles obtained by dividing a given angle by means of the interior bisector. This is the classical definition of the bisector.

30. PROPOSITION. The congruence of angles is an equivalence relation.

The proof is obvious and we leave it as an exercise for the reader.

31. PROPOSITION. Suppose that we have an angle $\widehat{hOk}$ in some plane α. Consider a plane α', a half-line $O'h'$ in it, and one (arbitrary) half-plane of α' whose origin is the line $O'h'$. Then there is a unique half-line $O'k'$, in the half-plane mentioned, such that $\widehat{h'O'k'} \equiv \widehat{hOk}$.

Indeed, take $A \in h$ and $B \in k$. Define $A' \in h'$ and B' in the half-plane of α' considered, such that $O'A' \equiv OA$ and $(O',A',B') \equiv (O,A,B)$. Then, $O'B'$ is the required half-line because the congruent transformation which sends (O,A,B) to (O',A',B') (Corollary 19) establishes the desired angle congruence. Uniqueness follows by a reductio ad absurdum.

32. PROPOSITION. Let (A,B,C) and (A',B',C') be two triples of noncollinear points such that $AB \equiv A'B'$, $AC \equiv A'C'$ and $\hat{A} \equiv \hat{A}'$. Then $\hat{B} \equiv \hat{B}'$ and $\hat{C} \equiv \hat{C}'$.

In fact, the congruent transformation which expresses $\hat{A} \equiv \hat{A}'$ sends the triple (A,B,C) to the triple (A',B',C') and implies thereby the relations $\hat{B} \equiv \hat{B}'$, $\hat{C} \equiv \hat{C}'$ too.

Now, it is possible to develop various other subjects as in classical books on elementary geometry, and we are quoting some of these subjects. (1) *Supplementary angles*. Two angles are supplementary angles if one of them is obtained from the other by the prolongation of one of its sides. In a broader sense, this name is also used if one of the angles is replaced by a congruent angle. One can prove that two angles are congruent iff their supplementary angles are congruent. (2) *Opposite angles*. The definition is classical, and opposite angles are always congruent. (3) *Right angles*. These are defined as having orthogonal sides, and all the right angles are congruent. (4) The *order relation* between angles. (5) *Acute*, *right* and *obtuse angles*. (6) *Algebraic operations* with angles. (7) *Rotations*. The rotations play an important role in the structure of general congruences. (8) For space one also has: *dihedral angles*, *rotations about an axis*, etc. [3, 6, 9, 10].

33. DEFINITION. Two triangles are called *congruent* if the corresponding sides of these triangles are congruent.

In other words, congruence of triangles is just congruence of respective triples of vertices and we could give the following equivalent definition: two triangles are equivalent if there is a congruent transformation which sends one of them to the other. Obviously, the triangle congruence is an equivalence relation. The following theorem contains the classical criteria for congruence of triangles:

34. THEOREM. 1° The corresponding angles of two congruent triangles are congruent. 2° If $AB \equiv A'B'$, $AC \equiv A'C'$ and $\hat{A} \equiv \hat{A}'$ the triangles ABC and $A'B'C'$ are congruent. 3° If $AB \equiv A'B'$, $\hat{A} \equiv \hat{A}'$ and $\hat{B} \equiv \hat{B}'$, the triangles ABC and $A'B'C'$ are congruent.

Indeed, 1° is a straightforward consequence of Definitions 29 and 33, 2° follows in the same manner as Proposition 32, and 3° follows by a reductio ad absurdum using Axiom V 6 and the uniqueness part of Proposition 31 (the details are left for the reader to work out).

An important auxiliary result is the *External Angle Theorem*:

35. THEOREM. An external angle of a triangle is greater than any of the internal nonadjacent angles of the triangle.

The names of internal, external and adjacent angles are the classical ones. As for the proof, let ABC be the triangle and Cx the prolongation of BC (Fig. 67), and suppose that we want to show $\widehat{ACx} > \hat{A}$. Take the middle M of AC, join B with M and extend this segment on the other side of M up to the point B' such that $B'M \equiv BM$. Using Theorem

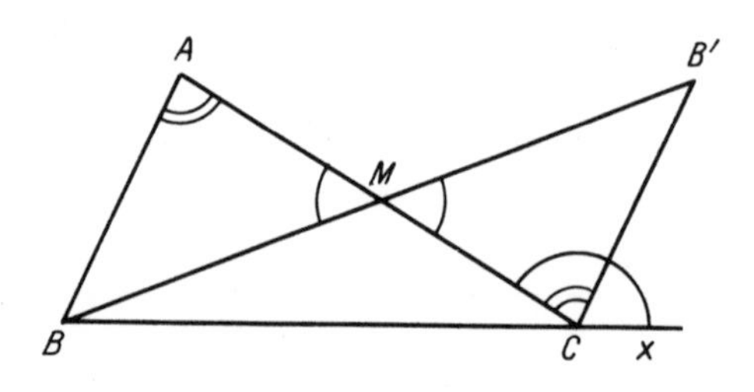

Fig. 67

34, 2°, we see that $ABM \equiv CB'M$, whence, by 34, 1°, $\hat{A} \equiv \widehat{ACB'} < \widehat{ACx}$, where the inequality follows since by the construction performed CB' is an interior half-line of $\widehat{ACx}$, q.e.d.

Using Theorem 35, we can also derive

36. THEOREM. Let ABC and $A'B'C'$ be two triangles such that $AB \equiv A'B'$, $\hat{A} \equiv \hat{A}'$ and $\hat{C} \equiv \hat{C}'$. Then $ABC \equiv A'B'C'$.

Indeed, it suffices to show that $AC \equiv A'C'$. In the contrary case, there is a point $C'' \neq C$ on the half-line AC such that $AC'' \equiv A'C'$. Then, in the half-plane of B with respect to AC, we have B'' such that $(A,B'',C'') \equiv (A',B',C')$ (Fig. 68). To avoid contradicting either Proposition 31 or Axiom V 5, we must have $B'' = B$ (see the third part of Fig. 68). Then, $\widehat{BC''A}$, which is external to the triangle BCC'', must be congruent with the internal nonadjacent angle $\hat{C}$ and Theorem 35 is contradicted. Hence, $C'' \neq C$ is impossible and Theorem 36 is proven.

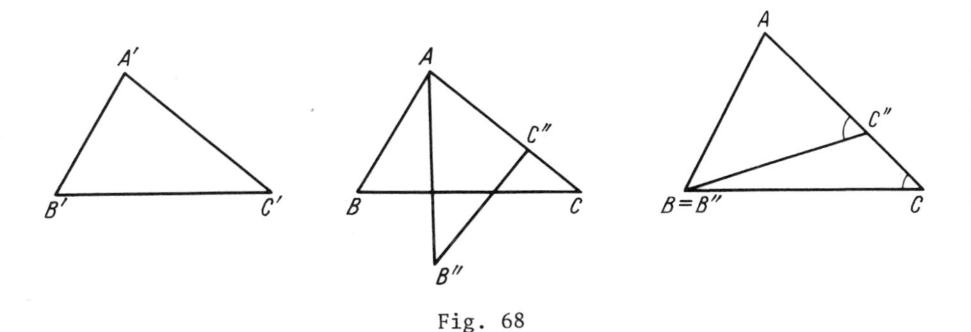

Fig. 68

Now, the classical theorems on congruence of right triangles, the comparison of sides and angles of congruent and noncongruent triangles, the triangular inequalities (stating that a side of a triangle is larger than the difference and smaller than the sum of the other two sides), etc., are available. It is also possible to study regular polygons and circles.

Finally, congruence theorems for the three-dimensional space are available as well, and it is noteworthy that there is no need of special spatial congruence axioms.

Before concluding this section, and for the sake of tradition, let us insert here the Hilbert formulation of the congruence axioms [9].

In [9], not only congruence of segments is a fundamental relation but also congruence of angles is such. Hilbert's congruence axioms were: V' 1 = V 5 without requiring the uniqueness of B', V' 2 = V $\bar{1}$, V' 3 = V 4, V' 4 = Proposition 31 together with the reflexivity of the congruence of angles, V' 5 = Proposition 32.

We have proven in this section that V implies V'. Conversely, it is possible to show that V' implies V. To do this, first, we see that the segment congruence is an equivalence relation; next, studying with V' congruence of triangles, we prove that it implies the congruence of corresponding angles. This in turn implies that angle congruence is an equivalence relation as well. Then the desired properties can be derived [6, 9].

Finally, before introducing the next section, let us prove

37. THEOREM. If two coplanar lines are cut by a third transversal line and enclose with it congruent internal alternate angles, the two lines are parallel.

The names of the angles made by a transversal with two coplanar lines are classical in elementary geometry. As for the proof of the theorem, if the two mentioned lines meet, they form with the transversal a triangle, and the congruent angles of the hypothesis would be, respectively, one external, the other internal and nonadjacent with respect to this triangle. This contradicts Theorem 35. Hence, the lines are necessarily parallel, q.e.d.

38. COROLLARY. Consider a line d and a point $A \bar{\in} d$. Then, there is at least one parallel line through A to d.

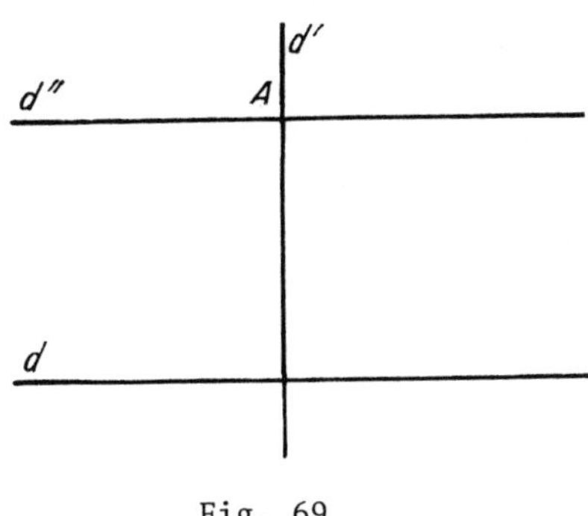

Fig. 69

Indeed (Fig. 69), consider the perpendicular line d' from A to d and also the line d'' of the plane (A,d) through A and orthogonal to d'. Then d'' and d' satisfy the hypotheses of Theorem 37 (because any two right angles are congruent). Hence $d'' \parallel d$, q.e.d.

Various other constructions of parallel lines are not excluded. Because of this, we shall give a special name to the parallel d'' obtained above, namely: the *canonical parallel* to d through A.

§2. EUCLIDEAN SPACES

a. Congruence and the Parallel Axiom

Now, we shall consider the geometric spaces which satisfy Axioms I, II, III' and V and which are special ordered affine spaces.

1. DEFINITION. The geometry defined by Axioms I, II, III, V, is called *Euclidean geometry* and the corresponding spaces are called *Euclidean spaces*.

We begin by mentioning a few straightforward consequences of the axioms. First, we note that, by Corollary 1.38, the parallel axiom may be given the weaker form II' (see 2, §4), requiring only the existence of at most one parallel to a line, through an exterior point. Actually, Corollary 1.38 excludes the Riemann geometry from the class of geometries considered here (i.e., assures the existence of the parallel) and II' excludes the possibility of more than one parallel to a line through an exterior point (i.e., excludes the Lobachevski geometry).

Another remark which is due at this point is that, in view of the parallel axiom II, we can now prove the converse of Theorem 1.37:

2. PROPOSITION. Two parallel lines enclose with any transversal congruent internal alternate angles.

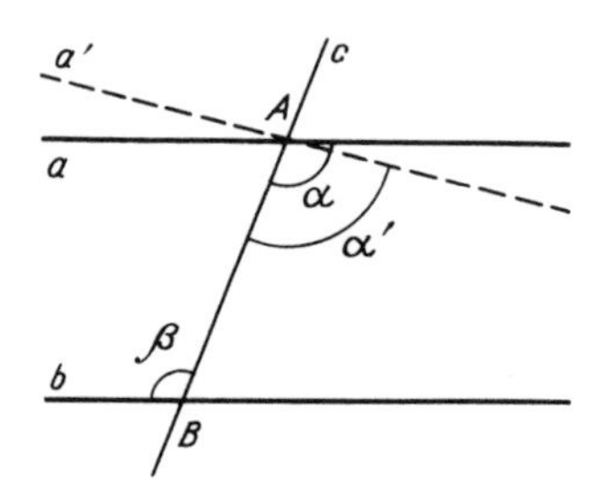

Fig. 70

In fact, let $a \parallel b$ and let the transversal c cut them respectively at A and B (Fig. 70). If $\hat{\alpha} \not\equiv \hat{\beta}$, we would have some other line a' through A such that $\hat{\alpha}' \equiv \hat{\beta}$ (Proposition 1.31), where $\hat{\alpha}', \hat{\beta}$ are the new internal alternate angles. According to Theorem 1.37, $a' \parallel b$ which is impossible, if $a \neq a'$, by Axiom II.

Of course, we also have the classical results for the other angles enclosed by parallel lines with a transversal.

If instead of adding Axiom II' to Axioms I, III', V, we add Proposition 2, then Axiom II' follows because, if in Fig. 70 we had $a \parallel b$, $a' \parallel b$, we would also have $\hat{\alpha} \equiv \hat{\beta}$, $\hat{\alpha}' \equiv \hat{\beta}$, in contradiction to Proposition 1.31. Hence there is no proof of Proposition 2 without the parallel axiom and this proposition is, actually, equivalent to the parallel axiom. Later on, we shall also see various other important equivalent propositions.

3. COROLLARY. Two coplanar lines orthogonal to a same third line of their plane are parallel. If a,b,c are three coplanar lines and $a \parallel b$, $c \perp a$, it follows that $c \perp b$.

Here, the first assertion follows from Theorem 1.37, and the second from Proposition 2 and from the congruence of any two right angles.

4. THEOREM. The sum of the angles of a triangle is equal to the sum of two right angles.

The proof can be easily recalled from elementary school books on geometry: consider through one of the vertices a parallel to the opposite side and use Proposition 2 to show that the sum of the angles of the triangle is equal to the sum of the angles on one side of a line. The details are left for the reader to work out. The significance of this theorem will be discussed, at greater length, at the appropriate section.

We can now produce a more precise *external angle theorem*:

5. COROLLARY. An external angle of a triangle is equal to the sum of the two internal nonadjacent angles of the triangle.

The proof is left to the reader to work out.

Finally, let us indicate the possibility of proving, at this stage, a lot of classical theorems about: parallelograms and other special quadrilaterals, the concurrence of perpendicular bisectors, of the altitudes, etc., of a triangle, spatial parallelism properties, and many others. We assume that these properties are known to the reader from the elementary books on geometry.

Let us also point out the relations between some notions which were introduced both by parallelism and by congruence. If $\overrightarrow{AB} \approx \overrightarrow{A'B'}$, in the sense of Definition 1.3.1, then we see by Proposition 2 that

$AB \equiv A'B'$. Hence, the same is valid for $\overrightarrow{AB} \sim \overrightarrow{A'B'}$, with AB and $A'B'$ on the same line, as well. Moreover, in this case we even have $\overrightarrow{AB} = \overrightarrow{A'B'}$ because, using an intermediary segment, the two segments are corresponding to each other by a product of two parallel projections, whence the orientation is preserved.

Conversely, if $\overrightarrow{AB} = \overrightarrow{A'B'}$ on a line (recall that equality means congruence and the same orientation) then let C be not on this line and D on the parallel through C to AB, in the same half-plane as B with respect to AC and such that $AB \equiv CD$. It easily follows that $\overrightarrow{CD} \approx \overrightarrow{AB}$ and $\overrightarrow{CD} \approx \overrightarrow{A'B'}$, whence $\overrightarrow{AB} \sim \overrightarrow{A'B'}$.

Consequently, the congruence vectors of Section 1 are identifiable with the usual "parallelism" vectors of Chapter 1. The same holds true for the congruence translations. It is obvious as well that the congruence middle coincides with the usual "parallelism" middle of Chapter 1.

b. The Scalar Field of an Euclidean Space

In the sequel, we want to study the scalar field of an Euclidean space. First, we have

6. THEOREM. The scalar field of an Euclidean space is commutative.

We know from Theorem 1. 3.20 that the assertion will be proven if we prove the Pappus-Pascal theorem for at least one pair of concurrent lines (this implies the general validity of the Pappus-Pascal theorem).

Let us take two orthogonal lines d,d' with the intersection point O and consider $A,B,C \in d$ and $A',B',C' \in d'$ such that $AB' \parallel BA'$ and $AC' \parallel CA'$ (Fig. 71). Take the perpendicular line through C' to AB'. By Corollary 3, this line is not parallel to d, whence it meets d at some point D. Then, C' is the intersection point of the altitudes of the triangle CDA' and we deduce that $CB' \perp DA'$. Next, since $AB' \parallel BA'$,

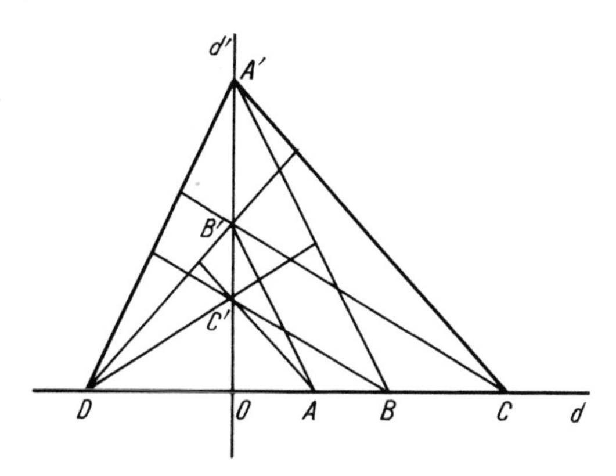

Fig. 71

C' is also the intersection point of the altitudes of the triangle BDA' which implies $BC' \perp DA'$. Then, using again Corollary 3, we get $BC' \parallel CB'$, which is just the Pappus-Pascal theorem.

Hence, Theorem 6 is proven.

Yet, this is not the only restriction on the scalar field of an Euclidean space. Other restrictions will be deduced by means of some new geometric results.

Let AB and CD $(C \neq D)$ be two arbitrary segments. Fix an origin O and some half-line with the origin O. Next construct the points P and Q characterized by $OP \equiv AB$, $OQ \equiv CD$ and by the fact that P and Q are on the fixed half-line. (This is made possible by Axiom V 5.).

It follows that we can associate with the pair (AB,CD) a scalar $k = (P,Q)_O$. This will be called the *ratio* of the segments AB and CD and it will be denoted by the *symbolic fraction* AB/CD.

The symbol AB/CD is well defined. Indeed, if we change the considered half-line through O, we have two other points P',Q' and $OP \equiv OP'$, $OQ \equiv OQ'$ (Fig. 72). Then, the bisector of $\widehat{POP'}$ will be orthogonal to both PP' and QQ', whence $PP' \parallel QQ'$ and $(P,Q)_O \sim (P',Q')_O$, i.e., we get the same scalar k as above.

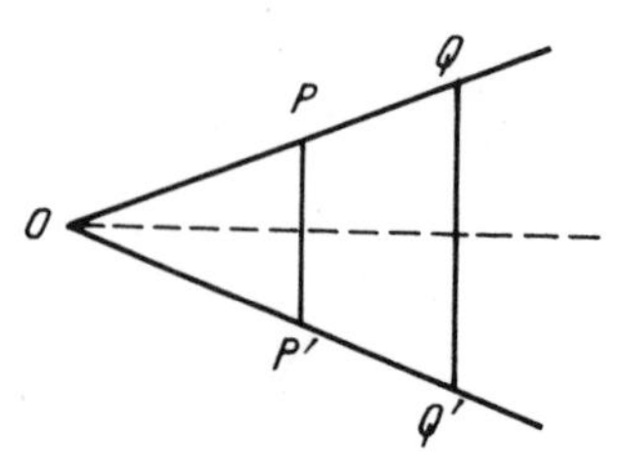

Fig. 72

If we change the origin O, we see, by choosing parallel half-lines that k is replaced by its image under the isomorphism of Proposition 1. 3.15. That is, in view of our conventions, we can say that the ratio remains unchanged.

Another, trivial, remark is that the ratio is unchanged if we replace the given segments by congruent ones.

The main properties of the ratios of segments are given by

7. PROPOSITION. The equality $AB/CD = EF/MN$ is true iff $AB/EF = CD/MN$ is true. The following relations hold for arbitrary segments

$$\frac{AB}{AB} = 1, \quad \frac{AB}{CD} \cdot \frac{CD}{EF} = \frac{AB}{EF}, \quad \frac{AB + A'B'}{CD} = \frac{AB}{CD} + \frac{A'B'}{CD}.$$

To arrive at the first assertion, put $OP \equiv AB$, $OQ \equiv CD$ and, on some other half-line through O, $OP' \equiv EF$, $OQ' \equiv MN$. Then, take on the first half-line $OR \equiv OP' \equiv EF$ and on the second $OR' \equiv OQ \equiv CD$ (Fig. 73).

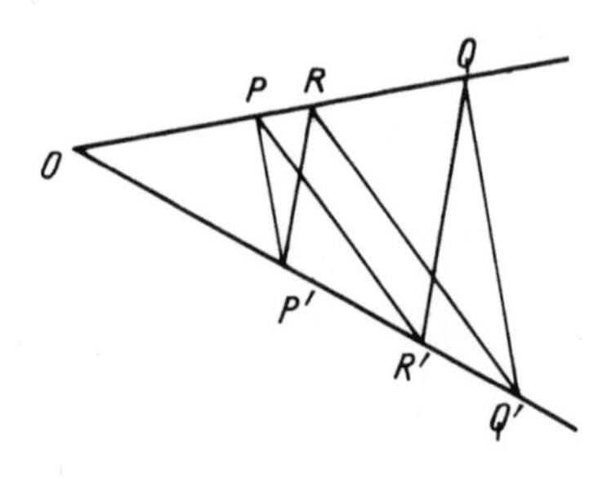

Fig. 73

As a result, we have (as in Fig. 72) $P'R \parallel QR'$ and the two ratio equalities mean respectively $PP' \parallel QQ'$ and $PR' \parallel RQ'$. The equivalence of these two parallelism relations follows from the Pappus-Pascal theorem, which holds now in light of Theorem 6.

As for the second part of the proposition, $AB/AB = 1$ is geometrically obvious and the second and third required relations are straightforward consequences of the definition of the scalar sum and product.

Furthermore, let us fix some segment UV ($U \neq V$) and call it the *unit segment*. Then, we can associate with an arbitrary segment AB the scalar $l(AB) = AB/UV$ which has the following obvious properties:

1° There are segments for which $l = 1$ (e.g., UV);

2° Two segments are congruent iff they have the same associated scalar;

3° $CD = AB + A'B'$ implies $l(CD) = l(AB) + l(A'B')$.

If K is the scalar field, $l(AB)$ is called the K-*length* of the segment AB or the K-distance from A to B. It is also worth noting that

$$\frac{AB}{CD} = \frac{l(AB)}{l(CD)},$$

which replaces the formal fractions denoting ratios by usual fractions in K, with usual operation rules. For example, from this remark we may derive that $AB/CD = EF/MN$ iff $l(AB)l(MN) = l(CD)l(EF)$.

If the unit segment is changed, all the K-length are multiplied by $UV/U'V'$, where $U'V'$ is the new unit segment.

A fourth interesting property of the function l is that: if k is an arbitrary positive scalar, there is some segment XY such that $l(XY) = k$. In fact, take the origin O and $OP \equiv UV$ on some half-line with the origin O. Next, put $k = (Q,P)_O$. Then OK is the required segment.

8. THEOREM (Thales). In a plane, a parallel to one of the sides of a triangle defines proportional segments on the other two sides.

Indeed, let ABC be our triangle and $PQ \parallel BC$, where $P \in AB$ and $Q \in AC$. If $A - P - B$, then $A - Q - C$ as well and taking A as the origin we clearly have $AP/AB = AQ/AC$. Going over to K-length and performing some classical computation with proportions, one easily derives $AP/PB = AQ/QC$, which is the desired result. If $A - B - P$, the proof is analogous. If $P - A - B$, then $Q - A - C$ and we shall apply the above proof to the points which are the reflections of P,Q with respect to A.

Theorem 8 is thereby completely proven.

Now, we can develop the theory of *similar triangles*.

9. DEFINITION. Two triangles ABC and $A'B'C'$ are called *similar* if their corresponding angles are respectively congruent (i.e., $\hat{A} \equiv \hat{A}'$, $\hat{B} \equiv \hat{B}'$, $\hat{C} \equiv \hat{C}'$).

As a matter of fact, because of Theorem 4, it suffices to verify only the congruence of two pairs of corresponding angles.

10. THEOREM. Two triangles are similar iff they are in one of the following situations: (a) they have one pair of corresponding congruent angles and the sides of those angles are proportional; (b) all their corresponding sides are proportional.

Indeed, let ABC, $A'B'C'$ be two similar triangles. Construct (Fig. 74) on the half-lines $A'B'$, $A'C'$ the points B'',C'' such that $AB \equiv A'B''$, $AC \equiv A'C''$. Then, the triangles ABC and $A'B''C''$ are congruent, whence $\hat{B}'' \equiv \hat{B} \equiv \hat{B}'$ and $B''C'' \parallel B'C'$. This implies $A'B''/A'B' = A'C''/A'C'$, i.e., $AB/A'B' = AC/A'C'$, thereby showing that (a) is a necessary condition. But, similarly, $AC/A'C' = BC/B'C'$, whence (b) is necessary as well.

Conversely, if (a) holds and $\hat{A} \equiv \hat{A}'$, then, after this construction we get $A'B''/A'B' = A'C''/A'C'$, whence $B''C'' \parallel B'C'$. This implies the required angle congruences and proves the similarity of ABC and $A'B'C'$.

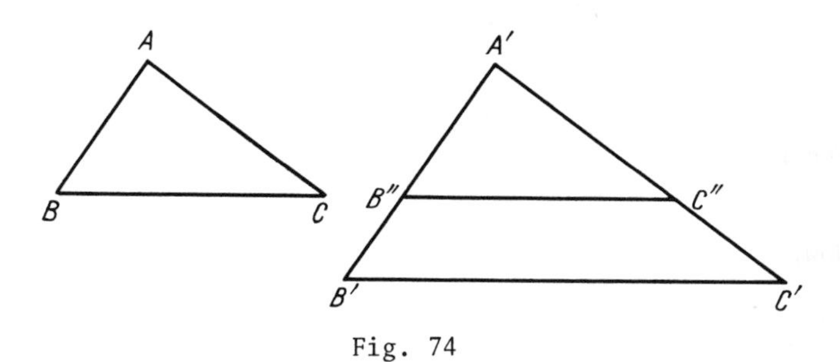

Fig. 74

If instead of (a), condition (b) is satisfied, we can derive through the same construction that the triangles $A'B''C''$ and $A'B'C'$ are similar which yields

$$\frac{B''C''}{B'C'} = \frac{A'B''}{A'B'} = \frac{AB}{A'B'} = \frac{BC}{B'C'} .$$

Therefore, $B''C'' \equiv BC$ and we have $ABC \equiv A'B''C''$, whence ABC and $A'B'C'$ are similar.

Hence, each of the properties (a), (b) is equivalent to the definition of similar triangles and (a), (b) are equivalent to each other as well.

In relation to the theory of similar triangles, we shall add the following remark [9]. In the above mentioned form, this theory was based on the consideration of the scalar field of the space and, particularly, in the case of the plane, on the Arguesian character of the plane. But, as a matter of fact, similarity does not require ratios, it merely requires proportions, i.e., equalities of ratios, and these use only parallelism (scalars are not actually needed). The results are similar (e.g., Theorem 10 holds), and it is simple to derive the plane special Desargues Theorem 1.2.13, without using three-dimensional space (The reader is asked to do this as an exercise.) That is, the congruence axioms imply the Arguesian character of the plane and *there are no non-Arguesian Euclidean planes*.

11. THEOREM (Pythagoras). In any right triangle ABC, where $\hat{A}$ is the right angle, the following relation holds

$$l^2(BC) = l^2(AB) + l^2(AC).$$

The proof uses similar triangles and it is well known from elementary books on geometry. We leave it as an exercise for the reader.

Now, we return to the scalar field of an Euclidean space.

12. DEFINITION. An ordered commutative field K is called an *Euclidean field* if for every two elements $a,b \in K$ there is some $c \in K$ such that $c^2 = a^2 + b^2$.

In other words, in an Euclidean field, the operation of square root of a sum of two squares is defined. Moreover, it is actually sufficient to ask for the existence of the elements $\sqrt{1+k^2}$ $(k \in K)$ only, because $\sqrt{a^2+b^2} = a\sqrt{1+(b/a)^2}$. A simple induction procedure shows the existence of the square root of an arbitrary finite sum of squares.

13. THEOREM. The scalar field of an Euclidean space is an Euclidean field.

Indeed, we already know that this is an ordered commutative field. If a,b are two arbitrary scalars, we may always construct such a right triangle ABC that $\hat{A}$ is a right angle and $l(AB) = a$, $l(AC) = b$. Then, by Pythagoras Theorem, we obtain $l(BC) = \sqrt{a^2+b^2}$, which proves the theorem.

c. Euclidean Structures and Quadratic Forms

The Pythagoras Theorem has important consequences.

In order to obtain them, let us begin with the following fundamental remarks. After choosing the unit segment of an Euclidean space, we can define the *length* of a vector by the K-length of the equipollent segments which define the vector. A vector of length 1 is called a *versor*. Every vector $\bar{v}$ is proportional to some versor, namely to $\bar{v}/|\bar{v}|$, where $|\bar{v}|$ is the notation for the length of the vector $\bar{v}$. Two vectors are called *orthogonal* if they have orthogonal supporting lines.

14. DEFINITION. A vector basis in an Euclidean space is an *orthonormal basis* if it consists of versors which are pairwise orthogonal. An affine frame is called *Euclidean* or *orthogonal cartesian* if its vector basis is orthonormal.

Let $\{O,\bar{e}_i\}$ $(i = 1,2,3)$ be an Euclidean frame and $\bar{v} = \sum_{i=1}^{3} x^i \bar{e}_i$ an arbitrary vector. Using the geometric definition of the coordinates x^i as shown in Fig. 30 of 1.§4. and the orthogonality of the frame, we get, by applying the Pythagoras Theorem twice, and using the notation of Fig. 30:

$$|\bar{v}| = l(OM) = [l^2(OM_1) + l^2(OM_2) + l^2(OM_3)]^{1/2}$$

Since $\bar{e}_i$ are versors, we get $l(OM_i) = x^i$, whence the formula which gives the analytical expresion of the length of a vector is

$$|\bar{v}| = \left[\sum_{i=1}^{3} (x^i)^2\right]^{1/2}. \tag{1}$$

Further, if $A(x^i)$, $B(y^i)$ are two arbitrary points, $l(AB)$ is the length of the vector $\overrightarrow{AB}$ constructed at the origin O and whose coordinates are $y^i - x^i$. This gives the analytic expression of the K-distance

$$l(AB) = \left[\sum_{i=1}^{3} (y^i - x^i)^2\right]^{1/2}. \tag{2}$$

Now, we can go to an arbitrary affine frame by a general linear transformation, and this leads obviously to some expression

$$|\bar{v}|^2 = \sum_{i,j=1}^{3} g_{ij}\, x^i x^j \qquad (g_{ij} = g_{ji}), \tag{3}$$

where the right-hand side is a nondegenerate and positive definite quadratic form on the vector space V of the vectors. (Such forms can be defined over an arbitrary ordered field just as over the real field.) A change of the unit segment multiplies the form (3) by some scalar factor. We have thus proven:

15. PROPOSITION. If, in an ordered affine space, a congruence relation of segments is defined, such that the space becomes Euclidean, then a canonical nondegenerate and positive definite quadratic form on the linear space V of the vectors can be introduced, and this form is determined up to an arbitrary positive scalar factor. This form defines the K-lengths and characterizes the segment congruences by the equality of the corresponding K-lengths.

These considerations also provide an analytic characterization of the group E of congruent transformations of the space. We know that these are affine transformations which send every segment to a congruent segment or, equivalently, preserve the K-lengths. Hence, these are transformations defined by formula (1) of 1.§5 where the associated centroaffine transformation preserves the quadratic form (3).

If the automorphism m of 1.§5.(1) is identical and if we use Euclidean frames, we get the usual orthogonality conditions [11] for the matrix p_j^i of the mentioned formula. The same conditions hold for the matrix expressing the transformation between two orthonormal frames.

So far, we have established the restrictions formulated by Theorem 13, for the scalar field of an Euclidean space. In the sequel, we prove that other restrictions do not exist since we have

16. THEOREM. If K is an arbitrary Euclidean field, there is a congruence relation for segments of the affine space over K, which provides it with the structure of an Euclidean space.

The affine space over K is the space constructed in Theorem 1. 4.6. Hence, the points are ordered triples of the form (x^i) $(i=1, 2,3;\ x^i \in K)$, and we can associate with every segment $A(x^i)$, $B(y^i)$ the *scalar characteristic* $l(AB)$ defined by formula (2) above. (We shall see later on that this is actually the K-length of AB).

Next, two segments AB and $A'B'$ will be called congruent if $l(AB)= l(A'B')$, and we shall denote this by $AB \equiv A'B'$. To prove the theorem, we must show that this relation satisfies Axioms V 1, V 4, V 5, V 6, and V 7. (Then, V 2 and V 3 follow by Proposition 1.16.)

V 1 is trivial: the segment congruence is an equivalence relation because the scalar equality is such.

For V 4, recall that the order of the points on a line is defined by considering the equation of that line under the form $\bar{r}=\bar{r}_0 + t\bar{v}$ and by taking the order of the values of the parameter t (Theorem 2. 3.9). This order remains unchanged if we replace $\bar{v}$ by $\rho\bar{v}$, where ρ is a positive element of K, and we can choose a factor ρ such that the new vector $\bar{v}$ has the property that the sum of the square of its coordinates equals 1 (i.e., we divide by the square root of this sum, which exists because K is an Euclidean field).

If we denote $A(\bar{r}_0)$, $B(\bar{r})$ and if $A<B$, then, using the previously chosen $\bar{v}$, we find $l(AB)=t$. Similarly, if $B<A$, we find $l(AB)=-t$. It follows that, if we have three points such that $A-B-C$ and if, this time, we put $B(\bar{r}_0)$ while A and C have t_1 and t_2 as the corresponding parameter values, then $l(AB)=-t_1$, $l(BC)=t_2$ and formula (2) gives $l(AC)=t_2-t_1$. Hence, $l(AC)=l(AB)+l(BC)$. Note that this is just the additivity property 3° of the K-lengths.

Now, finally, if $A-B-C$ and $A'-B'-C'$, then, by the above result, $l(AC)=l(AB)+l(BC)$, $l(B'C')=l(A'B')+l(B'C')$. Suppose that the hypotheses of Axiom V 4 hold. This means $l(AB)=l(A'B')$ and $l(BC)=l(B'C')$. But then $l(AC)=l(A'C')$ follow, i.e., $AC\equiv A'C'$ and Axiom V 4 is verified.

Take now Axiom V 5. If we use the notation from the formulation of that axiom, and if the line s has the equation $\bar{r} = \bar{r}_0 + t\bar{v}$, where $\bar{r}_0$ is the radius vector of A' while $\bar{v}$ is normalized as in the proof of V 4, then we have exactly two points B_1' and B_2' such that $AB \equiv AB_i'$ $(i = 1,2)$, namely, $\bar{r}_0 \pm l(AB)\bar{v}$. Since clearly A' is between these two points, only one of them is on the desired half-line.

The computations needed for the verification of V 6 and V 7 are lengthy computations of classical analytic geometry. Due to this fact, we shall only sketch them here.

First, the points of a plane may be represented under the vector form $\bar{r} = \bar{r}_0 + t_1\bar{u} + t_2\bar{v}$, where $\bar{u}$ and $\bar{v}$ can be chosen such that for each of them the sum of the squares of the coordinates be 1 and the sum of the products of the corresponding coordinates be zero.

Then, all the formulas for points of the plane have the same form as in classical geometry of the plane, where t_1 and t_2 are cartesian coordinates.

Now, if we use the notation of the formulation of Axiom V 6, we see that the required point C' must be at the intersection of the circles which have, respectively, center A',B' and radius $l(AC)$, $l(BC)$. These two circles will have exactly two intersection points, because the triangle inequalities in ABC show that the "distance" of the centers of the two circles is greater than the difference and smaller than the sum of the radii. Moreover, the line $A'B'$ cuts the interior of the segment joining the two intersection points mentioned. (All this is just like in the classical analytic geometry.). Hence, only one such intersection point is in the desired half-plane s, and Axiom V 6 is satisfied.

Finally, for Axiom V 7, using the notation from the formulation of that axiom, it is clear that $l(CP)$ is some function of the scalar characteristic of the other segments of the configuration. Next, $l(C'P')$ will be the same function of the corresponding segments of the second configuration, i.e., it will be the same function with the same values of the arguments. Hence $l(CP) = l(C'P')$ and Axiom V 7 is satisfied as well.

This completes the proof of Theorem 16.

Let us make some additional comments which relate to this theorem.

First, choose the unit segment OA, such that $O(0,0,0)$, $A(u^i)$ $(i = 1,2,3)$ and $\sum_{i=1}^{3} (u_i)^2 = 1$. Then, if we take $P(x^i)$, $Q(y^i)$ and use the definition of the K-length, we see that this is just the factor k appearing in the relations $y^i - x^i = ku^i$ which expresses the collinearity of $\overrightarrow{PQ}$ with some unit segment. By a simple computation, we also see that the scalar characteristic of the segment PQ is the same k. It follows that the scalar characteristics are just the K-lengths of the segments, as stated before.

Next, the proof of Theorem 16 actually shows that, whenever we attach to the affine space over an Euclidean field a nondegenerate and positive definite quadratic form on the corresponding vector space, this can be used as a scalar characteristic, and we can define a congruence relation for segments which makes the space an Euclidean space. Then, the square of the length of a vector is just the value of the quadratic form for that vector. The multiplication of the quadratic form by a positive scalar does not change the congruence relation. Combining these considerations with Proposition 15 we get

> 17. THEOREM. For an affine geometry over an Euclidean field, the definition of a congruence relation satisfying the fifth group of axioms is equivalent with the definition of a nondegenerate and positive quadratic form on the space of the vectors of that geometry, this form being determined only up to multiplication with an arbitrary positive scalar factor.

This theorem suggests an important generalization of Euclidean geometry consisting of the consideration of an affine geometry and an attached quadratic form on the corresponding space of vectors [1, 13, 14]. This form will define *lengths of vectors* and *scalar products of vectors*.

By using Theorem 17, we can introduce Euclidean geometry into the Erlangen Program. In fact, Theorem 17 shows that we must use as the basic group of this geometry the affine subgroups which multiplies by a positive scalar factor the associated quadratic form. Indeed, these transformations preserve the congruence relations, i.e., preserve the Euclidean structure of the space. Such transformations are called *similarities*, and Euclidean geometry can be thought of as the geometry of the group of similarities.

However, let us note that it is possible to prove that any similarity is the product of an homothety and a congruence transformation. This explains the essential role played by the group of the congruences in Euclidean geometry. Geometrically, it is more important to preserve not only the congruence relations but the K-lengths as well. Then, we have to take the Euclidean group E as the basic group of the space and the usual expositions of the Erlangen Program are defining the geometry of E as *Euclidean geometry*.

Finally, Theorem 17 shows that two Euclidean spaces with the same scalar field are isomorphic from the viewpoint of Euclidean gemetry. Indeed, we get an isomorphism of two such spaces by identifying points which have the same coordinates with respect to Euclidean frames. Then, the quadratic forms characterizing the congruence have both the same, canonical, expression. That is, there is essentially one Euclidean geometry over such a field. This remark together with Theorems 13 and 16 show that we have actually determined all Euclidean spaces, up to an isomorphism. Namely, these are the spaces over the Euclidean fields.

Furthermore, Theorem 16 allows for discussing the consistency of Euclidean geometry. Indeed, for instance, the field R of the real numbers is an Euclidean field. Therefore, the real affine space may be made into a Euclidean space whence: *Euclidean geometry and the field R of the real numbers are simultaneously consistent.*

Note that, for the same purpose, the rational field Q is not sufficient because it is not Euclidean (e.g., $\sqrt{1 + 1}$ does not exist in Q). This remark shows the independence of the group of congruence axioms of the axioms of affine geometry and, actually, on the axioms of the ordered Archimedean affine Pappian geometry.

However, R is not the only Euclidean field. For instance, we can consider the field Ω obtained by starting with the real number 1 and considering the results of an arbitrary finite sequence of operations of addition, subtraction, multiplication, division, and square roots of the form $\sqrt{1+a^2}$. This is clearly the smallest Euclidean subfield of R and it is not the whole of R because it consists only of algebraic numbers.

Another example of an Euclidean field is the field k (R,id.) of the formal power series, with real coefficients, defined in Chapter 2. We leave it to the reader to verify that, in this field, $\sqrt{1+\alpha^2} = \beta$ can be defined by considering indeterminate coefficients for β and, next, by finding these coefficients step by step such that $\beta^2 = 1 + \alpha^2$ is satisfied. Hence, the corresponding affine space can be made into an Euclidean space.

d. Real Euclidean Space

By now we have determined all the Euclidean geometries by Theorem 16. Furthermore, it is natural to consider those geometries which satisfy continuity axioms as well.

From Theorem 16 and Theorem 2.4.22, we see that the Archimedean Euclidean geometries are just the geometries over the Euclidean fields of real numbers. These are extensions of the field Ω above.

But we can see that the Cantor axiom remains an independent axiom even after the introduction of the congruence axioms. Indeed, in the contrary case the scalar field would necessarily be the real field. However, this is not true since, for instance, the geometry over Ω is both Euclidean and Archimedean but, as we already remarked, Ω is strictly included in R.

For an Archimedean Euclidean geometry, the K-lengths are real numbers and are simply called *lengths*. Hence the length is a function Segm $\longrightarrow R_+$, where R_+ is the set of non-negative real numbers (while Segm is the set of segments), which has the properties 1°, 2°, 3°, of the K-lengths. Length is a *measure* of the segments.

We can also define a *measure of angles*. Indeed, if $\bar{u}(u^i)$ and $\bar{v}(v^i)$ are two vectors, and if their coordinates are with respect to an orthonormal basis, then the expression

$$\frac{\sum_{i=1}^{3} u^i v^i}{\left[\sum_{i=1}^{3} (u^i)^2\right]^{\frac{1}{2}} \left[\sum_{i=1}^{3} (v^i)^2\right]^{\frac{1}{2}}} \tag{4}$$

does not depend on the choice of the orthonormal basis and, because of the famous Schwartz-Cauchy-Buniakowsky inequality, it has absolute value ≤ 1. Hence, this expression may be put in the form of $\cos\varphi$ and the absolute value of the argument φ, between 0 and π, will be, by definition, the measure of the unoriented angle between the two vectors.

The unit angle for this measure is called a *radian* and it is characterized by the fact that the measure of a right angle is $\pi/2$ radians.

Of course we can change the unit angle and take for instance the right angle as unit. That is (this in an interesting feature), while there is no canonical unit segment, there is a canonical unit angle. Further, we may define the measure of the angle of two half-lines to be the measure of the angle of two arbitrary vectors of these half-lines.

If we add the Cantor axiom then, as it follows from 2.§4, we are characterizing just the Euclidean geometry over the field of real numbers *which actually is the mathematical model of physical space*. This model is now determined up to an isomorphism.

Interestingly enough, Archimedes' axiom remains independent of Cantor's axiom in the Euclidean case as well. Indeed, the field

$k(R,\text{id.})$ is a Euclidean field (see the discussion above), which offers a non-Archimedean Euclidean geometry but satisfies the Cantor axiom (see 2.§4).

e. Absolute Geometry

We would like to conclude this section with an examination of yet another aspect related to the continuity axioms. Namely, the axioms of Archimedes and Cantor were formulated in 2.§4 in a manner which uses the affine structure of the space (see Proposition 2.§4.21 and Cantor's geometric axiom). Yet, both axioms together, are equivalent with the Dedekind Principle (Axiom IV) which is meaningful also in spaces satisfying only Axioms I and III'. Therefore, it is of interest to study the continuity properties of the spaces based on Axioms I, III', and V, i.e., to study geometries satisfying Axioms I, III', V, IV. Because all the common parts of Euclidean and Lobachevski geometries are included here, we call this *absolute geometry*. So far we have examined the consequences of axioms I, III', V, which are a part of absolute geometry. Now we shall explore the contribution added by Axiom IV.

First, once again, the Dedekind Principle can be replaced by a pair of axioms which, as a matter of fact, are just the axioms of Archimedes and Cantor formulated in terms of the congruence relation. To be more exact, these axioms are equivalent to the axioms of Archimedes and Cantor in any Euclidean space.

We denote these new axioms by IV'' and formulate them in the following manner:

IV'' 1. (Axiom of Archimedes). For any two segments $AB (A \neq B)$ and CD, there exists a natural number n such that $nAB > CD$.

IV'' 2. (Axiom of Cantor). Let $\{A_nB_n\}$ $(n = 0,1,2,\dots)$ be a sequence of strictly including segments such that, for every segment PQ, there is some rank n_0 with the property that $n > n_0$ implies $A_n B_n < PQ$. Then, there is a unique point M which is interior to all the segments of this sequence.

The equivalence of IV" 1 to IV' 1 and of IV"2 to IV' 2, if the parallel axiom holds, follows from the fact that, in this case, affine equipollence coincides with congruence equipollence (i.e., equality of oriented segments). As for the equivalence of IV to (IV" 1, IV" 2), it follows, mutatis mutandis, like Theorem 2.4.20 (i.e., by translating the corresponding proof into the language of the geometry on an ordered line). We leave this lengthy but not actually new proof to the reader to work out.

In this manner, we are led to examine absolute geometry in two steps: first, we shall add the axiom of Archimedes to I, III, V, and later on we shall add the Cantor Axiom too. The important results which we intend to prove here are the definition of coordinates (without parallelism) and the definition of a measure of segments.

18. DEFINITION. By a *coordinate system* on the oriented line d, we understand a function $x : d \to R$ with the following properties: (a) 0 and 1 belong to the range of this function; (b) x is a monotonical increasing function; (c) for two oriented segments of d, we have $\overrightarrow{PQ} = \overrightarrow{P'Q'}$ iff

$$x(Q) - x(P) = x(Q') - x(P').$$

From (a) and (b) it follows that there must be two well defined points O and E such that $O < E$ and $x(O) = 0$, $x(E) = 1$. These are called the *origin* and the *unit point* of the system x respectively, and, together with the orientation of d, they define a *frame* of d. For an arbitrary point $P \in d$, $x(P)$ is called the *abscissa* or the *coordinate* of P.

19. THEOREM. If Axioms I, III', V, and IV" 1 hold, then every frame of d defines a unique associated coordinate system on this line.

The following proof is due to H. Poincaré.

Consider the oriented line d and two points $O < E$ of d. Then take some point $P \geq O$ of d. We may construct an increasing sequence of points $\{P_n\}$ which is uniquely defined by the relations $P_0 = O$, $P_n P_{n+1} \equiv OP$. The axiom of Archimedes assures that for every index n there is a natural number φ_n which is the greatest one satisfying

$$\varphi_n OE \leq OP_n < (\varphi_n + 1)OE. \tag{5}$$

Multiplying (5) by $m > O$ and noting that $mOP_n = OP_{mn}$, we get

$$m\varphi_n OE \leq OP_{mn} < m(\varphi_n + 1)OE,$$

which yields, by the definition of φ_n:

$$m\varphi_n \leq \varphi_{mn} < m(\varphi_n + 1),$$

whence, by dividing by mn

$$\frac{\varphi_n}{n} \leq \frac{\varphi_{mn}}{mn} < \frac{\varphi_n}{n} + \frac{1}{n}. \tag{6}$$

Further, we derive from (6)

$$O \leq \frac{\varphi_{mn}}{mn} - \frac{\varphi_n}{n} < \frac{1}{n}$$

and by changing the role of m and n

$$O \leq \frac{\varphi_{mn}}{mn} - \frac{\varphi_m}{m} < \frac{1}{m}.$$

Now, a simple computation leads us to

$$\left| \frac{\varphi_m}{m} - \frac{\varphi_n}{n} \right| < \frac{1}{N} \qquad (N = \min\,(m,n)),$$

which justifies the existence of the number

$$x(P) = \lim_{n \to \infty} \frac{\varphi_n}{n} \tag{7}$$

and, by definition, we call this number the coordinate of P.

Next, for a point $P \leq O$ we shall define the coordinate $x(P) = -x(P')$, where $P' \geq O$ and $OP \equiv OP'$, i.e., P' is the symmetric point of P with respect to O.

In this manner, we get a function $x : d \to R$, and we shall show that it satisfies conditions (a), (b), and (c) of Definition 18.

First, if $P = O$ we have $\varphi_n = 0$ and if $P = E$ we have $\varphi_n = n$ for every n, whence $x(O) = 0$ and $x(E) = 1$. Therefore, property (a) holds and O, E are respectively the origin and the unit point.

Next, if $P \neq O$ then, by examining either P or its reflection by O, we see that we must have some $\varphi_{n_0} \neq 0$. By (6), this implies that φ_{mn_0}/mn_0 is a nondecreasing subsequence of $\{\varphi_n/n\}$ whose limit cannot be zero. Hence $x(P) \neq 0$, i.e., O is the only point sent to 0 by the defined function x.

All the results needed in this proof can be deduced from the following fact, which we shall prove now: if P,Q,R are three points of d, the relation $\overrightarrow{OR} = \overrightarrow{PQ}$ holds iff $x(R) = x(Q) - x(P)$.

It suffices to prove this for three points which are all $\geq O$ because the other cases are reducible to this one by taking corresponding symmetric points.

If $\overrightarrow{OR} = \overrightarrow{PQ}$, then $\overrightarrow{OP} + \overrightarrow{OR} = \overrightarrow{OQ}$, whence $\overrightarrow{OP} + \overrightarrow{OR} = \overrightarrow{OQ}$ and also $\overrightarrow{OP_n} + \overrightarrow{OR_n} = \overrightarrow{OQ_n}$, where P_n, Q_n, R_n are the sequences of points associated to P,Q,R.

The inequalities characterizing the sequences $\varphi_n^P, \varphi_n^Q, \varphi_n^R$ are

$$\varphi_n^P OE \leq OP_n < (\varphi_n^P + 1)OE,$$

$$\varphi_n^Q OE \leq OQ_n < (\varphi_n^Q + 1)OE,$$

$$\varphi_n^R OE \leq OR_n < (\varphi_n^R + 1)OE.$$

Adding the first and the third relation and comparing the result with the second one, we get in view of the definition of the numbers φ_n

$$\varphi_n^P + \varphi_n^R \le \varphi_n^Q < \varphi_n^P + \varphi_n^R + 2.$$

Dividing by n and taking the limit, we get

$$x(P) + x(R) = x(Q),$$

which is just the relation stated.

Conversely, if this relation holds and if $\overrightarrow{OR} \neq \overrightarrow{PQ}$, then there is some point S such that $\overrightarrow{OS} = \overrightarrow{PQ}$. It follows, by the direct part, that $x(S) = x(Q) - x(P) = x(R)$, and taking the point T defined by $\overrightarrow{OT} = \overrightarrow{RS}$ we obtain, using again the direct part, that $x(T) = x(S) - x(R) = 0$. But this contradicts $S \neq O$, which proves the required result.

We see now that x is a monotonical increasing function since, if $P < Q$, there is a point $S > O$ such that $\overrightarrow{OS} = \overrightarrow{PQ}$, whence $x(Q) - x(P) = x(S) > 0$.

Finally, an equality of the form $\overrightarrow{PQ} = \overrightarrow{P'Q'}$ is equivalent with the existence of a point S such that both segments are equal to $\overrightarrow{OS}$. Then, the equality is equivalent to $x(Q') - x(P') = x(Q) - x(P)$ $(=x(S))$, q.e.d.

Thus we have proven the existence of the announced coordinate system.

As for the uniqueness, let $y : d \to R$ be any coordinate system such that $y(O) = 0$ and $y(E) = 1$. Using property (c) of the coordinate system, it follows recursively that, for $P > O$ and for the sequence of points P_n as above, we must have $y(P_n) = ny(P)$. Then, from the monotony of the function y and by using the ends of the segments used in formula (5), we obtain

$$\varphi_n \le ny(P) < \varphi_n + 1,$$

whence, dividing by n and taking the limit, we derive $x(P) = y(P)$. If $P < O$, we get the same result by using the symmetric point of P with respect to O.

Theorem 19 is thereby completely proven.

Note that, when we also assume the parallel axiom and take an affine frame with origin O, with $\overrightarrow{OE}$ as one basic vector, then we have on d only one nonidentical zero affine coordinate which satisfies properties (a), (b), (c) of Definition 18 and is a real number (because the scalar field is Archimedean). Hence, this must be the abscissa of Theorem 19, in view of the uniqueness result mentioned above. That is, in the Euclidean case the coordinates of Definition 18 are affine coordinates.

Using the preceding uniqueness result, the reader is asked to show that, by a change of the frame, and eventually of the orientation of d, the abscissas are changed by a formula of the type $x' = ax + b$ $(a, b \in R)$. Indeed, it will be shown that, when a and b are conveniently chosen, x' has the properties of Definition 18 with respect to the new frame. (And this is true in general and not only in the affine case.)

20. THEOREM. If $x : d \to R$ is a coordinate system, $x(d)$ is everywhere dense in R.

Indeed, we must show (see §2 of the Introduction) that, for every $\varepsilon > 0$ and for every real number r, the interval $(r - \varepsilon, r + \varepsilon)$ contains values of $x(d)$.

Take first $r = 0$ and take the points $O < X_1 < \ldots < X_n < E$ such that $1/n < \varepsilon$. If we had $nOX_1 > OE$, $nX_i X_{i+1} > OE$ $(i = 1, \ldots, n - 1)$ and $nX_n E > OE$, we would obtain, by adding all these relations, $nOE > (n + 1)OE$. This contradiction shows that at least for one of the segments, say s, we must have $ns \leq OE$. Then, for the point $S > O$ for which $OS = s$, $x(S) \leq 1/n < \varepsilon$.

Now, take an arbitrary $r \in R$. Using the axiom of Archimedes for real numbers, we derive the existence of a natural number N such that

$$(N - 1)x(S) \leq r - \varepsilon < Nx(S),$$

whence

$$(r - \varepsilon) < Nx(S) < r + \varepsilon.$$

But $Nx(S) = x(S_N)$, where S_N is the corresponding point of the sequence S_n attached to S as in the proof of Theorem 19. That is, $Nx(S) \in x(d)$, which completes the proof of Theorem 20.

Theorem 20 clarifies the order structure of the line in absolute geometry without Cantor's axiom. Namely, the line is similar to an everywhere dense subset of the real field R. The congruence structure is clarified by property (c) of the coordinate systems.

We can also introduce a measure of the segments, which is characterized by the properties of the K-lengths.

21. DEFINITION. A *measure of segments* is a map $l : \text{Segm} \to R_+$ of the set of the segments to the set R_+ of real positive numbers, which satisfies the following conditions: (a) the number 1 belongs to the range of l; (b) congruent segments have the same image by l; (c) l is an additive function with respect to the sum of the segments.

If we have such a function l, any segment sent by l to 1 is called a *unit segment* and $l(AB)$ is called the *length* of the segment AB or the *distance* from A to B.

22. THEOREM. If a unit segment is fixed, there is a corresponding uniquely defined measure of the segments of the space.

In view of condition (b) of Definition 21, it suffices to define only the lengths of the segments of some fixed oriented line d. Take the points $O < E$ on d such that OE is congruent with the given unit segment, and define the coordinate system $x : d \to R$ associated with the frame (O,E). Next, if $A, B \in d$, put

$$l(AB) = |x(B) - x(A)|. \tag{8}$$

Then, the conditions of Definition 21 are trivially satisfied and this proves the existence part of Theorem 22. Moreover, we get the following "geometric interpretation" of the abscissa

$$x(P) = \begin{cases} l(OP) & \text{if } P \geq 0, \\ -l(OP) & \text{if } P \leq 0. \end{cases} \tag{9}$$

Further, if l is a segment measure and $l(OE) = 1$, the formulas (9) give a function $x : d \to R$, which satisfies (8) and is a coordinate system with origin O and unit point E. (The proof of this fact is left as an exercise for the reader.) Then, the uniqueness of the measure follows from the uniqueness of the mentioned coordinate system. Theorem 22 is thereby proven.

From the uniqueness part of Theorem 22, we can prove that a change of the unit segment implies the multiplication of all the lengths by $1/c$, where c is the length of the new unit segment with respect to the initial measure. In fact, by such a multiplication we obtain a map satisfying the conditions of Definition 21, with respect to the new unit segment.

Formula (8) and the fact that in the Euclidean case the abscissa is an affine coordinate imply that, in this case, the length given by Theorem 22 coincides with the length defined earlier in this section.

Note that it is possible to prove also an existence theorem for a measure of angles, based on Axioms I, III', V and IV''1.

Now, if Axiom IV'' 2 is added our results may be completed by

23. THEOREM. For every coordinate system $x : d \to R$, $x(d) = R$. Every positive real number is the length of a segment.

Indeed, let r be an arbitrary real number, consider then the following subsets of d

$$I = \{M/M \in d \text{ and } x(M) \leq r\}, \quad S = \{N/N \in d \text{ and } x(N) > r\}.$$

These are easily seen to define a Dedekind cut on d and, because IV'' 1 and IV'' 2 imply the Dedekind Principle, we have a corresponding defining point P. From Theorem 20, we see that the relations $x(P) < r$ and $x(P) > r$ are contradictory. Hence, we must have $x(P) = r$, which proves the first assertion of the theorem.

As for the second assertion, if $r > 0$, we have, by formula (8), $l(OP) = r$ for $x(P) = r$, q.e.d.

In other words, in the absolute geometry the lines are similar to the set R of real numbers and the congruence structure is given by the usual metric of R.

We can now define coordinates in the plane and in space by the following procedure: For a plane, fix some oriented line d in the plane and a frame (O,E) on d. Next, denote by + one of the two half-planes with origin d and by - the other half-plane. If P is any point of the plane, take $PP' \perp d$ and define the abscissa of P' on d with respect to the frame of d as abscissa of P. Next, define the *ordinate* of P'' as the length of the segment PP', with respect to the unit segment OE, with the sign + if P is in the half-plane + and with the sign - if P belongs to the half-plane -. The abscissa and the ordinata of P are the *coordinates* of P in the plane and, obviously, if we know these coordinates, the point P is well defined.

In space, fix some plane and define coordinates in this plane as above. Then, define a half-space + and a half-space - (the choice is arbitrary between the two half-spaces whose origin is the given plane). Finally, the coordinates of a point P are defined to be, first, the plane coordinates of the orthogonal projection P' of P on the considered plane, and then the length of $P'P$ with respect to the unit segment of that plane and with ± according to the sign of the half-space of P. Of course, if P belongs to the plane, the third coordinate of P is zero.

In the Euclidean case these coordinates coincide with the classical coordinates, with respect to a Euclidean frame.

Thus we have obtained some main results of absolute geometry.

§3. A SHORT HISTORY OF THE PARALLEL AXIOM

a. Historical Discussion

The main aim of this book, i.e., the mathematical construction of real Euclidean space, has been achieved. As we have seen, a very important role has been played by the parallel axiom and, even in the case of absolute geometry, a complete geometrical study is impossible without discussing the axiom of the parallels as well. This is logically explained by a fact which will be proven later, namely: that *the parallel axiom is independent of all the other axioms of absolute geometry*.

Before proceeding with the proof of the above mentioned result (the actual proof of which will follow in the next section), it is very interesting to make a quick survey of the history of this problem of independence; a problem which was in the limelight of geometry for over two thousand years and whose solution was a decisive step in the development of modern mathematics.

This survey will also provide us with many remarkable geometrical results consisting of propositions which are equivalent to the Euclidean parallel axiom, and have mathematical as well as historical importance.

We have already said (Introduction, §2) that elementary geometry began in the ancient era and that it flourished specially in ancient Greece. It is there that geometrical results were developed and systematized and the first axiomatic exposition of geometry appeared. All of these facts are contained in one of the most famous books which ever appeared: the *Elements* by Euclid (3d century B.C.). This book introduces the axiomatic method into science.

In the *Elements* geometry is developed starting with some number of descriptive definitions of what today we call fundamental notions, and with a number of postulates and axioms. This is followed by geometrical theorems which are deduced by proofs. The postulates and axioms are considered as intuitive truths.

One of the postulates, the fifth (or the eleventh axiom in other editions of the *Elements*) states that: *whenever a transversal line cuts two other lines and encloses with them internal angles in one half-plane such that the sum of these angles is less than the sum of two right angles, then the two lines have an intersection point belonging to the half-plane mentioned.*

Of course, this proposition pertains to plane geometry and it is this the original (fifth) parallel postulate (or axiom) of Euclid. We have

1. PROPOSITION. The fifth postulate of Euclid is equivalent to the parallel axiom II, provided that Axioms I, III' and V hold.

Indeed, if II holds, and if a,b are two coplanar lines cut by a transversal c and enclosing in one half-plane of c internal angles with the sum $< 2\,r$ (r is the notation which we shall use for a right angle), then a,b are not parallel because of Proposition 2.2. Hence, a and b have an intersection point C, and this cannot be in the half-plane where the sum of the internal angles is $> 2\,r$, because, in the opposite case, a triangle ABC, as shown in Fig. 75, appears for which, by Theorem 1.35 about the external angle, we had $\hat{A}_1 > \hat{B}_2$, $\hat{B}_1 > \hat{A}_2$ and $\hat{A}_1 + \hat{B}_1 > \hat{A}_2 + \hat{B}_2 > 2\,r$, in contradiction to the hypothesis $\hat{A}_1 + \hat{B}_1 < 2\,r$.

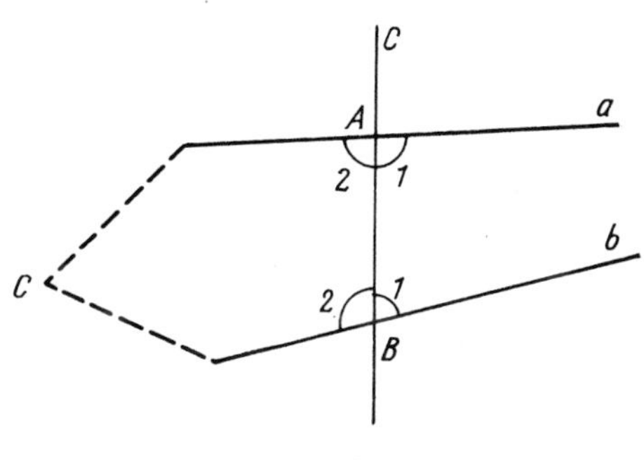

Fig. 75

Conversely, the fifth postulate immediately implies Proposition 2.2, which in turn, implies Axiom II.

The other axioms and postulates of Euclid (which, generally, are older than the *Elements*) describe what we call today the absolute geometry. This was an incomplete description and, with the time, new postulates (continuity, order) were added, but, essentially, one may say that the axioms of the absolute goemetry had been stated.

From the very beginning, because the axioms were regarded as intuitive truths, the parallel axiom was considered as critical. Indeed, it is intuitively difficult to say whether two lines for which the sum of the angles mentioned in the postulate is close enough to $2r$ actually meet.

In fact, the problem of the parallel lines was approached critically even before Euclid. For instance, Aristotle shows that, from the logical viewpoint, a special axiom is needed in the theory of parallel lines and, probably, formulated himself such an axiom.

This development led to the question of whether it is possible to prove the fifth postulate from the other postulates and axioms, and thereby make it into a theorem. This is exactly the problem of the independence of the parallel axiom with respect to the other axioms of absolute geometry.

The solution to this problem was given only two thousands years later, in the 19th century, after a very long period of research. Many mathematicians thought that they had proven the postulate but had actually proven equivalent propositions.

Naturally, the problem was very difficult. Indeed, because of that attitude towards the axiomatic method, which considered axioms as intuitive truths about some intuitively known objects, only the idea of the dependence of the parallel axiom was taken into account, while the independence of this axiom seemed contradictory.

Moreover, the mathematicians had no complete list of axioms of absolute geometry.

Thus, the solution to the problem became possible only when a qualitatively new kind of mathematical thinking was developed, and in itself this solution contributed much to such a development. Ultimately, it led to the modern axiomatic method, to complete and rigorous foundations of geometry, and one can say that it marked the beginning of modern formal mathematical thinking.

Many people think that Euclid himself believed that the fifth postulate should be a theorem and tried to prove it. Their argument is that the fifth postulate was avoided as much as possible in the *Elements*. Even so, it is more interesting to note that Euclid stated it as a postulate and did not present some false proof as many other mathematicians did later.

Beginning with Euclid and a long period afterwards, geometers tried hard to prove the parallel postulate or at least to reduce it to intuitively more acceptable propositions.

Apparently, Archimedes had written a paper entitled "On Parallel Lines," which was not found but was mentioned by medieval mathematicians, in which parallel lines were defined either by the property that the distance from the points of one line to the second line is constant or as the trajectories of two points of a rigid body in a rectilinear uniform movement. Both definitions keep reappearing in the later writings of some mathematicians who used the works of Archimedes (Posidonius, first century B.C., Aganis, 5th century A.D., Ibn Kora, 9th century A.D.). We shall see later that the existence of such lines is actually equivalent to the Euclidean parallel axiom. Proclus (5th century A.D.) proves the fifth postulate under the hypothesis that the distance from the points of a line to a parallel line is bounded.

Other important research relates the parallel axiom with different kinds of quadrangles and with the sum of angles of a quadrangle and of a triangle.

Thus, Ibn al Chaisam (Alchazen) (10th century A.D.) considers quadrangles with three right angles and shows that the parallel axiom is equivalent to the fact that the fourth angle is right as well.

The Persian poet and mathematician Omar Khayyam (11th century A.D.) uses quadrangles with two right angles having a common side (the *basis*) and the other two sides equal. Then, the perpendicular bisector of the basis is a symmetry axis (why?) and the other two angles of the quadrangle, called *upper angles* are equal. The parallel axiom is equivalent to the fact that these upper angles are right as well.

Consequently, three hypotheses are available in absolute geometry for the discussed angles of both types of quadrangles mentioned above: *the right angle hypothesis, the acute angle hypothesis* and *the obtuse angle hypothesis*. The Euclidean parallel axiom is equivalent to the right angle hypothesis. In absolute geometry, one can show that the obtuse angle hypothesis is contradictory and one can try to show that the acute angle hypothesis is contradictory as well. But this was impossible in view of the independence of the parallel axiom which was proven later.

The existence of the above types of quadrangles, i.e., of the rectangles, means the existence of quadrangles for which the sum of the angles is $4r$. If a diagonal of such a quadrangle is used, then one gets triangles whose angles have the sum $2r$, and this relates the problem of the parallel axiom to the sum of the angles of a triangle.

Later, similar developments were also considered by Western European geometers, and the first of them was, probably, the Jewish scientist Levi ben Gershon (Gersonides), in the 14th century.

In the 18th century, Gerolamo Saccheri of Italy studied the parallel lines by using quadrangles like those of Omar Khayyam, and Lambert used quadrangles with three right angles.

Both of them obtained very important results. It is interesting to note that in the *Theory of Parallel Lines* of Lambert (1766) one finds no claim as to the proof of the parallel postulate. Moreover, the possibility of producing a non-Euclidean geometry on an *imaginary sphere* is suggested. (Actually, the Lobachevski geometry holds on a sphere of imaginary radius if this sphere is considered in a space whose *metric* is defined by a quadratic form, whose canonical form has one negative term.)

Finally the last comprehensive and important researches which aimed to prove the Euclidean parallel axiom were those of Legendre (1752-1833).

Legendre studied the sum of angles of a triangle, and tried, by reductio ad absurdum, to eliminate the possibility of this sum being $> 2r$ or $< 2r$. If this happens, then the sum must be $2r$ and this can be seen to be equivalent to the Euclidean parallel axiom. In fact, Legendre shows that the hypothesis of a sum of angles $> 2\,r$ is contradictory but, when getting the same conclusion for the sum $< 2\,r$, he admits that by every interior point of an angle there is some transversal of both sides of the angle. As a matter of fact, and as it was later proven, this property is equivalent to the Euclidean parallel axiom. (A similar error was made by Simplicius, 6th century, Al Djahari, 9th century, and Samarkandi, 13th century.)

Then, followed the discovery of the non-Euclidean geometry, which was the unexpected solution to the parallel problem and which we shall describe later on. This solution was neither understood nor widely recognized at first and various false proofs of the Euclidean parallel postulate continued to appear, but they are of no interest at this time.

b. Mathematical Discussion

Since such results as those of Legendre, Saccheri, Lambert, etc. are not only of historical but also of mathematical interest, we shall consider part of them here. We begin with

> 2. PROPOSITION. If all the axioms of absolute geometry hold, the parallel axiom II is equivalent with the fact that the sum of the angles of every triangle is equal to $2\,r$.

In proving this result, we shall use the properties of the angle measure which, as we know, can be obtained as a consequence of the continuity axioms. We use also isosceles triangles and their properties. The corresponding definition is classical and the properties are easy consequences of the congruence axioms.

We already know, by Theorem 2.4, that Axiom II implies the result.

Now, suppose that the sum of the angles of an arbitrary triangle is $2\,r$ and take a line d and a point $M \bar{\in} d$. Take $MP \perp d$, $d' \perp MP$ through M and in the plane (M,d), and a new line d'' through M in the same

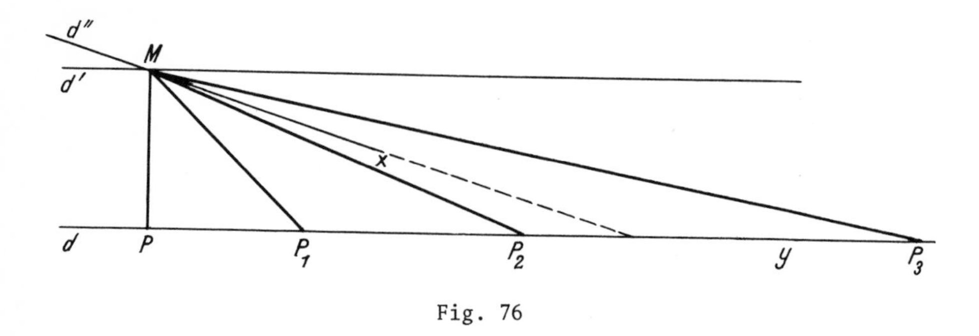

Fig. 76

plane. $d' \parallel d$ and d' is what we called the canonical parallel through M to d. We have to show that the relation $d'' \parallel d$ is impossible.

Suppose that Mx is the half-line of d'' such that $\widehat{PMx} < 1\,\mathrm{r}$. Let s be the half-plane with origin MP which contains Mx, and let Py be the half-line of d situated in s (Fig. 76). Then, let P_1 be a point of Py such that $PP_1 \equiv PM$ and construct a sequence of points P_i such that $P_{i-1} - P_i - P_{i+1}$ and $P_{i-1}P_i \equiv P_{i-1}M$.

The triangle $\widehat{MPP_1}$ is right and isosceles, whence, by the hypothesis, $PMP_1 = 1/2\,\mathrm{r}$. Similarly, all the triangles $P_{i-1}MP_i$ are isosceles and, using Corollary 2.5, $\widehat{P_{i-1}MP_i} = 1/2^i\,\mathrm{r}$. Then, by summing, we get

$$\widehat{PMP_n} = 1\,\mathrm{r} - \frac{1}{2^n}\,\mathrm{r}.$$

By taking a sufficiently large n, we have $\widehat{PMP_n} > \widehat{PMx}$, and, by Proposition 2.1.15, since Px is interior to the angle PMP_n, Px cuts PP_n. It follows that d'' is not parallel to d and this completes the proof of Proposition 2.

Proposition 2 suggests defining the following notion of absolute geometry: let $S(ABC)$ denote the sum of angles of the triangle ABC; put

$$D(ABC) = 2\,\mathrm{r} - S(ABC);$$

this is called the *defect* of the triangle ABC.

3. PROPOSITION. The defect of a triangle is additive with respect to the decomposition of the corresponding triangular plate into a sum of triangular plates.

The proof is by induction on the number of terms. If the triangle is decomposed into two plates the result follows by a straightforward computation (an exercise for the reader to carry out), and for the case of n terms we shall get a decomposition into two plates, each of them decomposed into less than n plates. Alternatively, a proof based on Euler's Theorem 2.2.15 can also be given.

4. PROPOSITION. In absolute geometry, we have for every triangle $D \geq 0$, i.e., $S \leq 2\,\mathrm{r}$.

In fact, take a triangle ABC (Fig. 77), and let $\widehat{ABx}$ be one of its external angles. Then, by Theorem 1.35, we have

$$\widehat{ABC} + \widehat{ACB} < \widehat{ABC} + \widehat{ABx} = 2\,\mathrm{r}.$$

Hence, the sum of two angles of a triangle is always less than $2\,\mathrm{r}$.

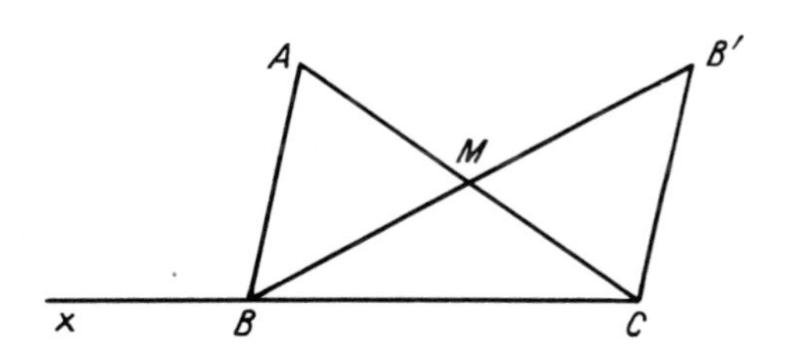

Fig. 77

Now, suppose $S(ABC) = 2\,\mathrm{r} + \varepsilon$, with $\varepsilon > 0$. Let M be the midpoint of AC and B' be the symmetric point of B with respect to M (Fig. 77). It easily follows that

$$S(ABC) = S(B'BC), \quad \widehat{ABB'} \equiv \widehat{BB'C},$$

which means that we have constructed a new triangle, $B'BC$, which has the same angle sum $2r+\varepsilon$ and which has an angle $\leq$ half of a well defined angle of the initial triangle ABC (because either $\widehat{B'BC}$ or $\widehat{BB'C}$ is less than $1/2\ \hat{B}$).

By repeating this procedure n times, we get a triangle PQR such that $S(PQR) = S(ABC)$ and, for instance, $\hat{P} \leq 1/2^n\ \hat{B}$. But, for large enough n, $1/2^n\ \hat{B} < \varepsilon$, whence $\hat{Q} + \hat{R} > 2r$ in contradiction to the result established about the sum of two angles of a triangle.

This proves Proposition 4.

5. COROLLARY. The Euclidean parallel axiom is equivalent with the fact that the defect of any triangle is zero. In the geometry of Lobachevski, there are triangles with a strictly positive defect.

However, we do not know yet that every triangle has a positive defect in the Lobachevski geometry. This will be seen to hold as a consequence of another important result:

6. PROPOSITION. If the axioms of absolute geometry hold, then the Euclidean parallel axiom is equivalent with the existence of at least one triangle for which the sum of angles is $2r$ (i.e., the defect vanishes).

Indeed, we shall show that, if the sum of angles of one triangle is $2r$, then the same is true for every triangle. The following proof follows from Legendre. A proof of this fact without the continuity axioms is also available, but the proof of Proposition 2 uses essentially the continuity axioms [10].

First, we note that, if we have a triangle with vanishing defect, we also have a right-angled triangle with vanishing defect. Indeed, using a conveniently chosen altitude of the given triangle, we get a decomposition into two right triangles and the sum of their defects is zero by Proposition 3. Because, by Proposition 4, every term of this sum is nonnegative, these terms are zero, which proves our assertion.

By the same decomposition, we can see that, if the defect of every right triangle is zero, then the defect of every triangle is zero as well.

Hence, what we actually have to show is that if a right triangle ABC with $\hat{A} = 1\,\mathrm{r}$ and $D(ABC) = 0$ exists, then, for every right triangle PQR ($\hat{P} = 1\,\mathrm{r}$), we have $D(PQR) = 0$.

Consider first the case when $AB \geq PQ$ and $AC \geq PR$ (Fig. 78). Then, take the points E and F on AB and BC respectively, such that $AE \equiv PQ$ and $AF \equiv PR$. The triangles AEF and PQR are congruent, hence they have the same defect. In view of Propositions 3, 4, we have

$$0 \leq D(PQR) = D(AEF) \leq D(AEF) + D(FEC) + D(CEB) = D(ABC) = 0,$$

whence $D(PQR) = 0$ as required.

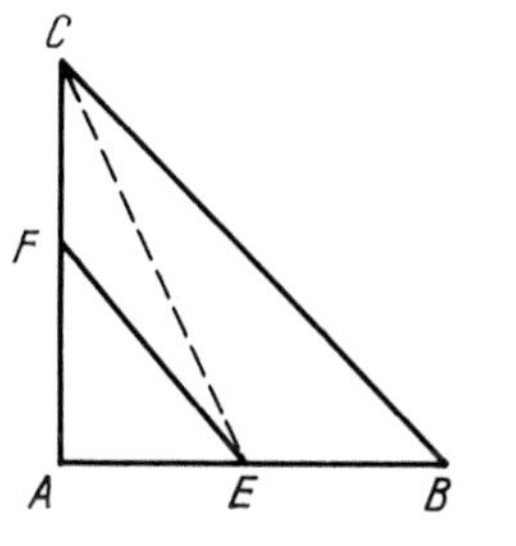

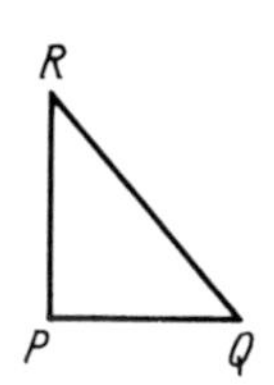

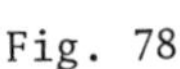

Fig. 78

Further, if at least one of the relations $AB \leq PQ$, $AC \leq PR$ holds, we shall construct from ABC a new right triangle, with vanishing defect and such that the sides of its right angle are greater than PQ, PR, and this will complete the proof of Proposition 6.

To do this, take again the points E, F on AB, AC respectively, such that $AE \equiv PQ$, $AF \equiv PR$ (Fig. 79). Then, define A_1 on the other side of BC than A and such that $CA_1 \equiv AB$ and $\widehat{A_1CB} \equiv \widehat{CBA}$. The triangles ABC and A_1CB are congruent, and ABA_1C is a rectangle. By a repeated application of this construction, we get the *pavement* of Fig. 79.

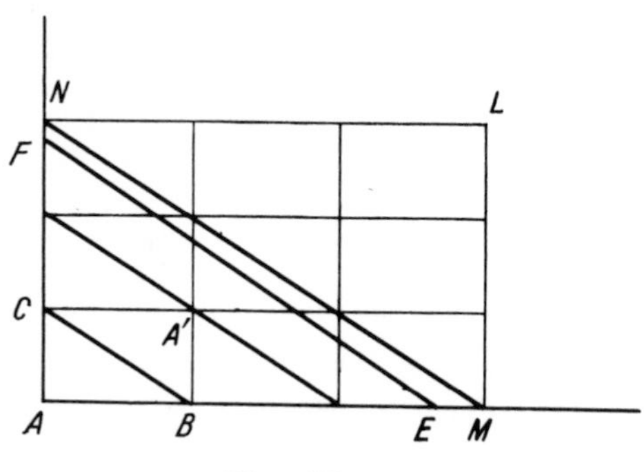

Fig. 79

Because of the Axiom of Archimedes, we reach some rectangle $AMLN$ such that $AM > AE$ and $AN > AF$.

We have now

$$S(AMLN) = S(AMN) + S(MLN) = 4\,r,$$

and, then $S(AMN) \leq 2\,r$, $S(MLN) \leq 2\,r$ (Proposition 4), imply $S(AMN) = 2\,r$. (Otherwise $S(MLN) > 2\,r$, which is impossible.)

Hence AMN is the required right triangle which has vanishing defect and sides of the right angle greater than those of ABC.

Proposition 6 is thereby proven.

Note that the last part of the proof holds for every quadrangle with the sum of angles equal to $4\,r$ which proves:

7. PROPOSITION. The Euclidean parallel axiom is equivalent, modulo the axioms of absolute geometry, with the existence of at least one quadrangle in which the sum of angles is equal to $4\,r$.

The expression "modulo the axioms of the absolute geometry" means: when these axioms hold.

We can also easily prove:

8. PROPOSITION. The Euclidean parallel axiom is equivalent, modulo the axioms of absolute geometry, with the fact that all triangles have the same sum of angles.

Indeed, in Euclidean geometry $S(ABC)$ is always $2\,\mathrm{r}$. Conversely, suppose $S(ABC) = \alpha$ for every triangle. Then, choose a point D such that $B - D - C$, and decompose ABC into two triangles by means of the segment AD. Write down in an explicite form $S(ABC) = \alpha$, $S(ABD) = \alpha$, $S(ADC) = \alpha$. Since $\widehat{BAD} + \widehat{DAC} = \widehat{BAC}$ and $\widehat{BDA} + \widehat{CDA} = 2\,\mathrm{r}$, one obtains easily $\alpha = 2\,\mathrm{r}$ which clearly implies Proposition 8.

As an outcome of the Legendre results we also have

9. PROPOSITION. The Euclidean parallel axiom is equivalent, modulo absolute geometry, with the existence of at least one angle such that, through every interior point of it there is a transversal, cutting the two sides of the angle.

First, in Euclidean geometry this property is valid for every angle. Indeed, if $\widehat{hOk}$ is some angle and M an interior point, then the parallels through M to Oh and Ok intersect Ok and Oh in P,N respectively (Fig. 80). Any transversal which is in the interiors of the angles $\widehat{NMx}$ and $\widehat{PMy}$ of Fig. 80 intersects both Oh and Ok.

Conversely, let $\widehat{hOk}$ be an angle which has the stated property (Fig. 81), and take $A \in Oh$, $B \in Ok$. Next, take O_1 on the other side of

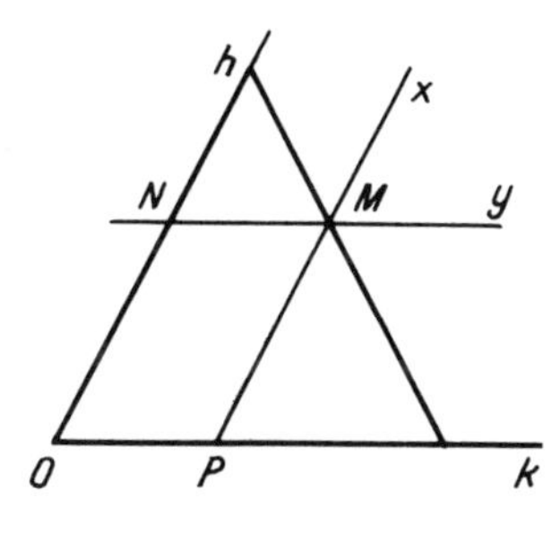

Fig. 80

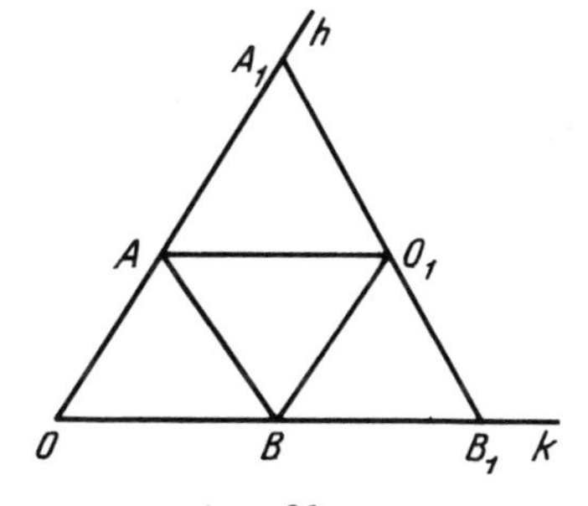

Fig. 81

AB than O and such that $\widehat{BAO_1} \equiv \widehat{ABO}$ and $AO_1 \equiv OB$. It is easy to see that O_1 is interior to the angle $\widehat{hOk}$ and that the triangles OAB and O_1BA are congruent. Hence there is a transversal A_1B_1 through O_1 such that $A_1 \in h$, $B_1 \in k$ and we have

$$D(A_1OB_1) = 2D(AOB) + D(AA_1O_1) + D(BB_1O_1) \geq 2D(AOB).$$

By repeating this construction, we get a triangle A_nOB_n such that

$$D(A_nOB_n) \geq 2^n D(AOB).$$

It follows that, if $D(AOB) \neq 0$, we can get triangles with arbitrary large defects. But this is impossible because, by the definition of the defect, it must be $\leq 2\,\mathrm{r}$.

Hence, necessarily $D(AOB) = 0$ and we complete the proof by quoting Proposition 6.

The same proof also yields

10. PROPOSITION. The Euclidean parallel axiom is equivalent, modulo absolute geometry, with the existence of one acute angle such that the perpendicular at any point of one side meets the second side of the angle.

In Euclidean geometry, this is true for every acute angle because Oh and $Px \perp Ok$ (Fig. 82) are coplanar lines which are met by the transversal Ok such that $\widehat{hOk} + \widehat{xPO} < 2\,\mathrm{r}$.

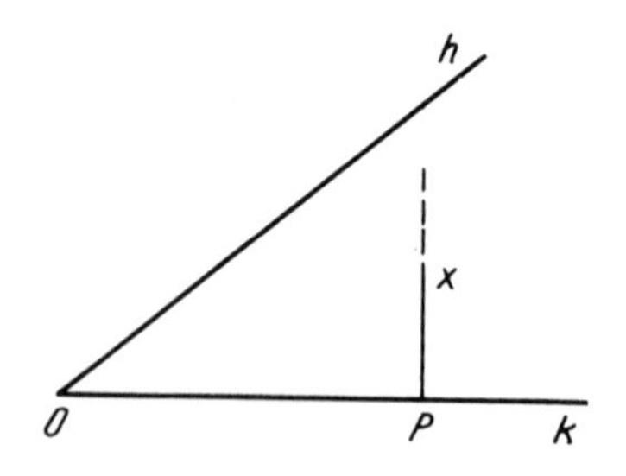

Fig. 82

The converse result follows just as in the case of Proposition 9 if we use as corresponding transversals the perpendicular from O_1 to Ok, and so on.

Another interesting result is

11. PROPOSITION (Wallis, 18th century). The Euclidean parallel axiom is equivalent, modulo absolute geometry, with the existence of two similar and noncongruent triangles.

The existence of such triangles in Euclidean geometry is an almost trivial fact: it suffices to take a triangle and to consider through a point of one of its sides a parallel line to another side. Then, the third side is intersected and we obtain a triangle which is similar and noncongruent to the initial one.

Conversely, suppose ABC and $A'B'C'$ are similar triangles, i.e., the corresponding angles are congruent, and suppose $AB < A'B'$. Then, consider the point B'' such that $A' - B'' - B'$ and $AB \equiv A'B''$ and, also, the angle $\widehat{A'B''x} \equiv \hat{B} \equiv \hat{B}'$ (Fig. 83). Then $Bx \parallel B'C'$ and, by Pasch's Axiom, Bx meets $A'C'$ in C''. The triangles ABC and $A'B''C''$ are congruent, whence $\hat{C} \equiv \hat{C}'' \equiv \hat{C}'$. This easily implies $S(B'C'C''B'') = 4\,r$, and the desired result follows by Proposition 7.

The same Proposition 7, implies the following results for the triple-right quadrangles of Al Chaisam-Lambert and for the double-right quadrangles of Omar Khayam-Saccheri:

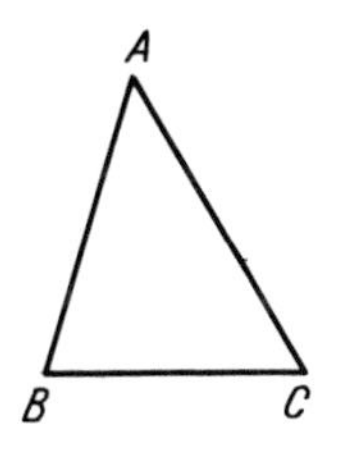

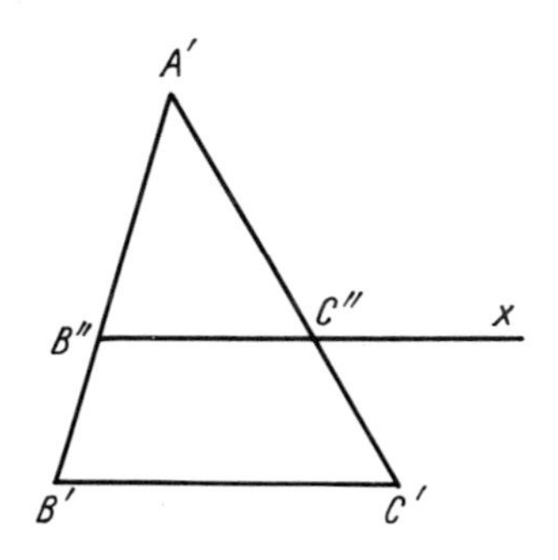

Fig. 83

12. PROPOSITION. The Euclidean parallel axiom is equivalent, modulo absolute geometry, with the existence of one Lambert quadrangle with the fourth right angle or with the existence of one Saccheri quadrangle with upper right angles.

Hence, Euclidean geometry actually corresponds to the *right angle hypothesis*. The *obtuse angle hypothesis* contradicts Proposition 4, and the *acute angle hypothesis* holds in the Lobachevski geometry.

We may now prove the results which relate parallelism with the distances from the points of one line to another line.

13. PROPOSITION. The Euclidean parallel axiom is equivalent, modulo absolute geometry, to each of the following assertions: (a) the locus of the points of a half-plane whose distances to the origin-line of the half-plane are constant is again a line; (b) the distances from the points of a line to a parallel line are bounded.

Both (a) and (b) are obvious in Euclidean geometry.

Conversely, suppose (a) holds. Let d be a line and P,Q,R three points in the same half-plane and at equal distances from d. By (a) these are collinear points. Take $PL, QM, RN \perp d$. The quadrangles $LMQP$, $MNRQ$, $LNRP$ are of the Saccheri type, whence $\widehat{LPQ} \equiv \widehat{MQP}$, $\widehat{MQR} \equiv \widehat{NRQ}$ and $\widehat{LPQ} \equiv \widehat{NRQ}$ (Fig. 84). It follows $\widehat{PQM} \equiv \widehat{MQR}$ and the upper angles of the three Saccheri quadrangles are right angles. Then, Proposition 12 implies the desired result.

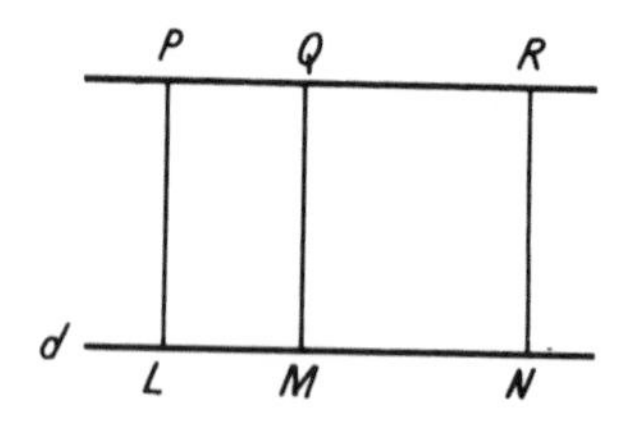

Fig. 84

To get the result when (b) holds, we shall prove that if the Euclidean parallel axiom is not true then the distances from the points of a line to a parallel line are unbounded.

Take $d' \parallel d$ and with nonconstant distances between them (the case of constant distances is settled by (a)). Take $A, B \in d'$, $AM \perp d$, $BN \perp d$ and suppose $AM > BN$ (Fig. 85). Denote $h = AM - BN$ and take E such that $M - E - A$ and $ME \equiv NB$. Then, $MNBE$ is a Saccheri quadrangle and, because the Euclidean parallel axiom is not true, $\widehat{MEB}$ is acute. Hence, by the External Angle Theorem, $\widehat{MAB}$ is also an acute angle and $\widehat{CAM}$ is obtuse.

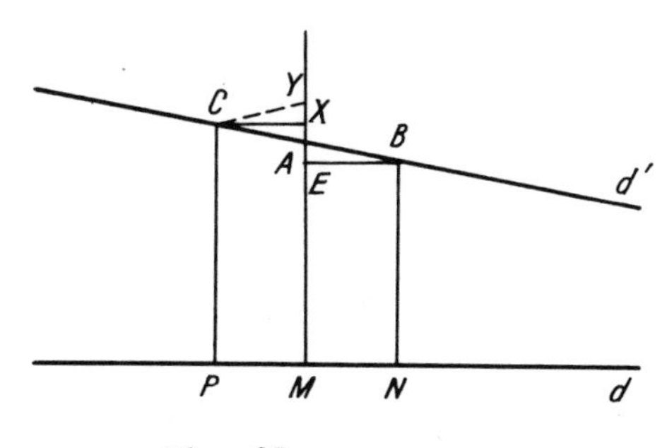

Fig. 85

Next, consider the order $A < B$ on d' and take $C < A$ such that $AC \equiv AB$, and $CP \perp d$. We note that $CP > AM$ since, in the opposite case, we would have some point X such that $M - X - A$ and $PC \equiv MX$, and $CPMX$ is a Saccheri quadrangle.

Then $\widehat{CXM}$ is acute and, at the same time, it is greater than the obtuse angle $\widehat{CAM}$, which is contradictory. Hence if for X one has $PC \equiv MX$, one also has $M - A - X$.

Let us also show that $AX \geq AE$. In the contrary case, there would be a point Y with $A - X - Y$ and $AY \equiv AE$. It follows that the triangles CAY and BAE are congruent, whence $\widehat{CYA} \equiv \widehat{AEB}$, and these are obtuse angles. And, because of the External Angle Theorem, the acute angle $\widehat{CXA}$ would be greater than the obtuse angle $\widehat{CYA}$, which is, of course, impossible.

It follows that $CP \geq BN + 2h$ and, by repeated application of the same procedure, we get points C_n on d' such that $C_nP_n \geq BN + 2^n h$, which implies the unboundedness result, q.e.d.

We have thus completed the mathematical examination of the propositions which were encountered in the short historical notes at the beginning of this section. Now, we shall proceed with a few other similar propositions of interest, which are equivalent to the Euclidean parallel axiom.

Let us give the name of a *midline* to a segment joining the midpoints of two sides of a triangle, while, in this case, the third side will be called the *basis*. We have:

14. PROPOSITION. The Euclidean parallel axiom is equivalent, modulo absolute geometry, with the property that the midline of a triangle is equal to one half of the corresponding basis.

This property of the midline in Euclidean geometry is classical and we leave its proof to the reader.

Suppose now that the midline is one half of the basis.

Consider a triangle ABC and let M,N be the middles of AB,AC respectively. Take $AL \perp MN$ and suppose that the triangle considered is such that $M - L - N$ (Fig. 86; clearly such triangles exist). Extend the line MN on both sides and take $PM \equiv ML$, $QN \equiv NL$ as in Fig. 86. Then, the triangles AML, BMP are congruent and ANL, CNQ are also congruent.

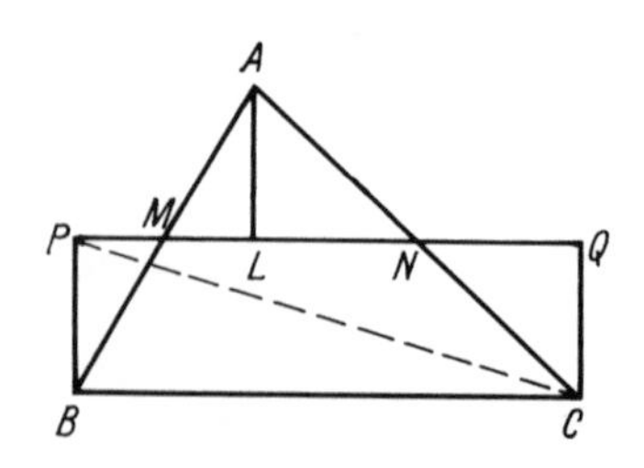

Fig. 86

It follows that $\widehat{BPM}$ and $\widehat{CQN}$ are right angles and $BP \equiv LA \equiv CQ$. Hence, $PQCB$ is a Saccheri quadrangle with the basis PQ and the *upper angles* $\hat{B}$ and $\hat{C}$. From the hypothesis $MN = 1/2\ BC$ we easily get $PQ \equiv BC$, whence we deduce that BCP and QPC are congruent triangles. Then, $\widehat{PBC} \equiv \widehat{CQP}$ and the upper angles of the mentioned Saccheri quadrangle are right angles.

This completes the proof of Proposition 14. In fact, we see that it suffices to require the property of the midline for only one triangle and one midline which are such that, in the notation of Fig. 86, the condition $M - L - N$ holds.

15. COROLLARY. Modulo absolute geometry, the Euclidean parallel axiom is equivalent with the Pythagoras Theorem.

A proof of the Pythagoras Theorem in Euclidean geometry was already mentioned (Theorem 2.11). Now, suppose that Pythagoras Theorem holds in any right triangle and take one such triangle ABC, $\hat{A} = 1\,r$. Let M,N be the midpoints of AB,AC respectively. Then

$$MN = \sqrt{AM^2 + AN^2} = \frac{1}{2}\sqrt{AB^2 + AC^2} = \frac{1}{2}BC$$

and the result follows by Proposition 14.

16. PROPOSITION. The parallel axiom II is equivalent, modulo the Axioms I, III, V, with the fact that two lines which are parallel with the same third line are parallel to each other.

The proof is left as an exercise for the reader to work out.

17. PROPOSITION (Farkas Bolyai). The parallel axiom II is equivalent, modulo the Axioms I, III, V, with the fact that any three noncollinear points belong to a circle.

Indeed, as we have already mentioned earlier, from I, II, III, V, we can deduce that the three perpendicular bisectors of a triangle are concurrent (if these were parallel, the vertices would be collinear). Clearly, their intersection point is the centre of the required circle.

Conversely, suppose circumscribed circles exist, and take a line a and a point $A \bar{\in} a$. Take $AP \perp a$ and $a' \perp AP$ through A, i.e., a' is the canonical parallel $a' \parallel a$. If $b \neq a'$ and $A \in b$ (Fig. 87), then take $Q \in AP$, Q' the symmetric point of Q with respect to a and Q'' the symmetric point of Q with respect to b. a and b will be perpendicular bisectors of the triangle $QQ'Q''$ and they must have a common point: the centre of the circle which contains Q, Q', Q''. Hence $b \parallel a$ cannot hold, whence the uniqueness of the parallel, q.e.d.

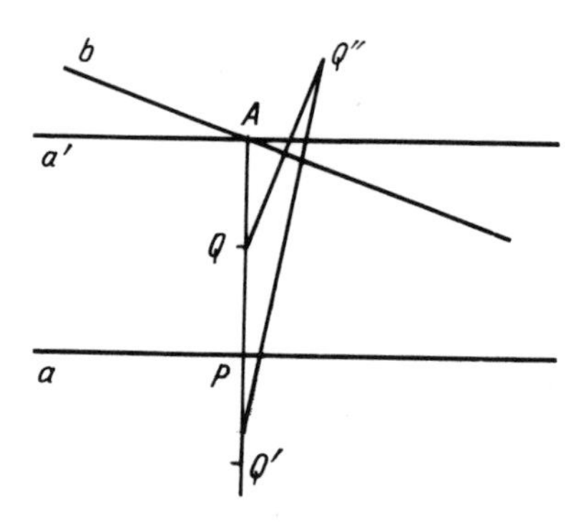

Fig. 87

Finally, let us indicate how the correct solution of the parallel problem has been obtained. The solution was found in the third decade of the 19th century, almost simultaneously and independently of each other by C.F. Gauss, N.I. Lobachevski and J. Bolyai. By a critical study of the unsuccessful proofs of Euclid's fifth postulate, they recognized that it cannot be proved and that non-Euclidean geometry is as consistent as Euclidean geometry.

It is probable that Gauss knew about these results even before 1820 but did not publish anything during his lifetime about the parallel problem. His ideas are known to us from letters and notices published much latter. Moreover, he did not encourage similar ideas of J. Bolyai, who had sent his obtained results to Gauss.

János Bolyai had also discovered non-Euclidean geometry and, in 1831 published his results in the form of an Appendix to the book on geometry written by his father, Farkas Bolyai. The results of J. Bolyai are most often developed as *absolute theorems*.

More inclusive results were established and published by N.I. Lobachevski, whose first communication on this subject dates back to 1826. In 1829, he published the first paper on non-Euclidean geometry and, afterwards, continued to work on the subject for the rest of his life. Lobachevski proves the main theorems of the geometry in which, through an exterior point of a line one has more than one parallel to the given line. He also proves essentially that the new geometry is as consistent as Euclidean geometry. This new geometry is called the *Geometry of Lobachevski*.

The solution to the problem of the parallel lines had a revolutionary character for the whole of mathematics.

For the content of the non-Euclidean geometries we refer the reader to [3, 5]. Here, we shall be interested in proving the consistency of these geometries, because this is equivalent to the independence of the parallel axiom with respect to the axioms of absolute geometry. The actual proof shall be provided in the next section.

§4. THE INDEPENDENCE OF THE PARALLEL AXIOM

There are many interesting models of Lobachevski's geometry, which prove both the consistency of this geometry and the independence of the parallel axiom of the axioms of the absolute geometry. These models can also be used for the study of non-Euclidean geometry. Here, we choose one of these models, which is simple enough and very useful, the *Poincaré model*, and we present it first for plane geometry.

Take a real Euclidean plane π, a line a of it and a half-plane s of π with origin a. The line a will be called the *ideal line* of the model, and s will be called the *upper half-plane*. Also, we shall add to π a *point at infinity* which will allow us to consider π as the plane of a complex variable z.

We call *h-point* every point of the upper half-plane and *h-line* every half-line and every semicircle which lie in the upper half-plane and are orthogonal to the ideal line. In the case of a semicircle, this means that its centre is on the ideal line. The points of the ideal line are not h-points. Hereafter, we shall use the name *semicircle* for both proper semicircles and half-lines (where the half-lines may be considered as semicircles with centre at infinity) and, if no explicit contrary assumption will be made, the centres of the semicircles will be on the absolute line and the half-lines will be orthogonal to the absolute line. Also, we denote by P the set of all h-points with the h-line-structure. We shall show that P admits the structure of a Lobachevski plane.

The verification of the incidence axioms I 1,2,3 is very simple and we leave it as an exercise for the reader.

Next, based on Euclidean geometry, it is clearly meaningful to speak of a betweenness relation for points of a semicircle, and this relation satisfies the order axioms III' 1,2,3. We can also verify that Pasch's Axiom holds.

Indeed, let ABC be an h-triangle and d an h-line which contains none of the points A,B,C (Fig. 88).

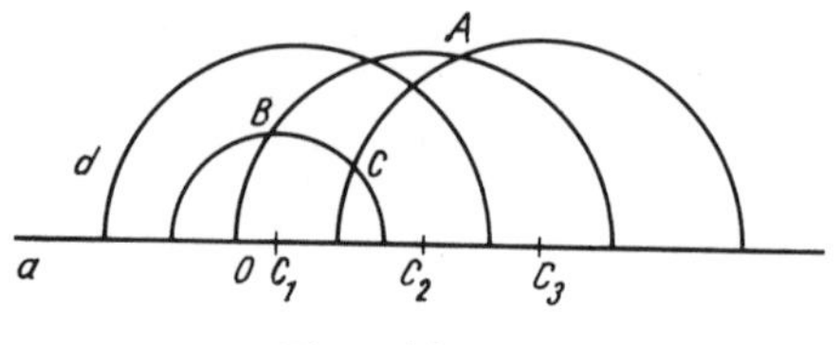

Fig. 88

Let $F(x,y) = 0$ be the equation of d with respect to some Euclidean frame, such that $F(x,y) < 0$ represent the interior of the circle d (or one of the half-planes of d if this is a half-line).

Denote by (x_i, y_i) $(i = 1,2,3)$, respectively, the coordinates of A, B, C and let $\alpha_i = F(x_i, y_i)$.

Now, if d meets the interior of the segment AB, we have $\alpha_1\alpha_2 < 0$, whence $(\alpha_1\alpha_3)\cdot(\alpha_2\alpha_3) = \alpha_1\alpha_2\alpha_3^2 < 0$, and either $\alpha_1\alpha_3 < 0$ or $\alpha_2\alpha_3 < 0$. It follows that d also intersects either AC or BC which is just the Pasch Axiom.

Now, if we project a semicircle from its centre onto a line parallel to the ideal line, we get a bijective correspondence which, obviously, is order preserving. It follows that an h-line is similar (from the order viewpoint) to a Euclidean line, whence the h-lines satisfy also the Dedekind Principle IV. Thus P is a model for the continuity axioms as well.

In order to prove that P is a model of absolute geometry, we must verify the congruence axioms, which is a more complicated task.

We have already mentioned that we shall use the interpretation of the plane π as the plane of the complex variable $z = x + iy$, by choosing an Euclidean frame of π and adding some new point $z = \infty$.

Then, if $M_i(z_i)$ $(i = 1,2,3,4)$ are four points of the plane π, we define their *cross-ratio* by the formula

$$(M_1, M_2; M_3, M_4) = \frac{z_3 - z_1}{z_3 - z_2} : \frac{z_4 - z_1}{z_4 - z_2}. \tag{1}$$

This quantity is a real number and has a geometric meaning if the points M_i belong to a line or to a circle. Namely, in the first case, one gets

$$(M_1, M_2; M_3, M_4) = \frac{l(M_1M_3)}{l(M_2M_3)} : \frac{l(M_1M_4)}{l(M_2M_4)},$$

where l denotes Euclidean length, and in the case of a circle one gets

$$(M_1, M_2; M_3, M_4) = \frac{\sin \widehat{M_1PM_3}}{\sin \widehat{M_2PM_3}} : \frac{\sin \widehat{M_1PM_4}}{\sin \widehat{M_2PM_4}},$$

where P is an arbitrary point of the respective circle.

If we call (M_1, M_2) and (M_3, M_4) the *pairs* of the cross-ratio, M_1, M_4 *extremal points* and M_2, M_3 *medium points*, we immediately get from (1) the following properties: (a) the cross-ratio remains unchanged if either the role of the two pairs or the order of the points of both pairs is changed; (b) by changing the order of the points of one pair only, the value k of the cross-ratio (1) changes to $1/k$; (c) by changing the position of either the two extremal points among them or of the two medium points, the cross-ratio k changes to $1 - k$.

It follows that we can always bring any one of the four points of a cross-ratio to the first position or to the last position by transformations which do not affect the value of the cross-ratio. Hence, from the 24 possible cross-ratios of four nonordered points only 6 can have distinct values and all these values are simple functions of one of them. If $(M_1, M_2; M_3, M_4) = -1$, the two pairs are called *harmonic pairs* and in this case only 3 of the 6 values are distinct.

If, for instance, $z_4 = \infty$, one has from (1)

$$(M_1, M_2; M_3, \infty) = \frac{z_3 - z_1}{z_3 - z_2}$$

and if, moreover, M_1, M_2, M_3 are collinear points, this is just the *simple ratio* of these three points. If another $z_i = \infty$, we bring it first to the last position and next calculate the cross-ratio.

If one knows three points and the value of the cross-ratio, the fourth point is well defined by formula (1).

For any five points the following *Möbius relation* holds

$$(M_1, M_2; M_3, M_5) = (M_1, M_2; M_3, M_4)(M_1, M_2; M_4, M_5) \tag{2}$$

By definition, if a cross-ratio is negative, we say that the two corresponding pairs are *separating* each other. For collinear or concyclic points this can be seen to hold iff the pairs are separating in the geometric meaning of this word.

Let us also recall that the *inversion* of *pole* O and *power* k is the (plane or space) transformation which sends any point $P \neq O$ to P' collinear with O and P and such that $OP \cdot OP' = k$ (we have here the product of lengths and the sign is fixed by means of an orientation), and sends O to ∞. In the complex representation, and using O as the origin, this transformation is given by $z' = k/\bar{z}$, where the bar denotes complex conjugation.

It follows by an easy computation that, by an inversion, a cross-ratio is sent to its complex-conjugate. Since in the particular case of collinear or cocyclic points the cross-ratio is a real number, it will actually be invariant by inversions.

Moreover, the inversions send circles not containing the pole to circles and circles through the pole to lines. Also, lines are sent to circles through the pole or to lines according to whether they do not or do contain the pole. Finally, the inversions are *conformal transformations*, i.e., angle-preserving.

The reader is asked to establish for himself all these properties of cross-ratios and inversions.

As a first application, let us show how we can use cross-ratios in order to characterize h-half-lines. Note that those are meaningful because P satisfies the order axioms.

Let d be an h-line, $A \in d$ and U,V the points where the semicircle d meets the absolute line (Fig. 89).

It is geometrically clear that the two h-half-lines with origin A on d are represented by the arcs AU and AV, and that a point B is

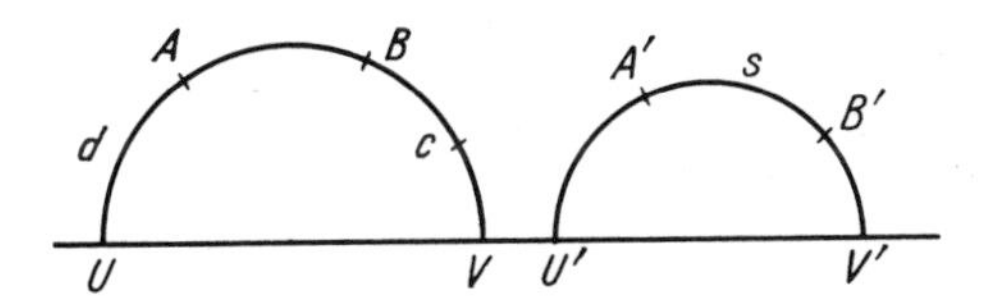

Fig. 89

on AV iff the pairs (U,B) and (A,V) separate each other. This means $(U,B;A,V) < 0$. But, a simple computation, using the above mentioned properties of the cross-ratios, shows that this condition is equivalent to

$$(U,V;A,B) < 1. \tag{3}$$

Similarly, the point B belongs to AU iff

$$(U,V;A,B) > 1. \tag{4}$$

Hence, these relations are the announced characterization of the h-half-lines. In the case when d is a half-line, we take $V = \infty$ and the same characterizations hold.

The h-half-planes defined by d are represented by the interior and the exterior regions of the proper semicircle $\underline{d}$ in the upper half-plane of the model or, respectively, by the two regions defined by a half-line d orthogonal to the ideal line.

Now, let A and B be two h-points and d of Fig. 89 the h-line joining them. We associate with the h-segment AB the real characteristic

$$h(AB) = R\,|\ln(U,V;A,B)|, \tag{5}$$

where $R > 0$ is some real number called the *curvature radius* of the Lobachevski plane. If $V = \infty$, formula (5) reduces itself to

$$h(A,B) = R\,\left|\ln \frac{AU}{BU}\right| .$$

$h(AB)$ is well defined because (U,V) and (A,B) are non-separating pairs, whence $(U,V;A,B) > 0$. From the properties of the cross-ratio, it follows that $h(AB) = h(BA)$.

Then, we define the congruence of two h-segments in P by $AB \equiv A'B'$ if $h(AB) = h(A'B')$ and we shall prove that, by this definition, all the congruence axioms V are satisfied in P.

Indeed, V 1 is trivial and V 2, V 3, are not interesting because they are dependent axioms.

Further, if we have some h-points $A - B - C$ (Fig. 89), it follows, in view of formulas (3), (4), that $(U,V;A,B)$ and $(U,V;B,C)$ are simultaneously <1 or >1, whence, by the Möbius relation (2) we easily deduce that

$$|\ln(U,V;A,C)| = |\ln(U,V;A,B)| + |\ln(U,V;B,C)|.$$

That is,

$$h(AC) = h(AB) + h(BC) \tag{6}$$

and we see that h has the additivity property of the segment measure. Now, Axiom V 4 is a trivial consequence of formula (6).

To verify Axiom V 5, we shall use the notation of the formulation of this axiom, i.e., we have to obtain a point B' such that

$$h(A'B') = h(AB) = a.$$

This means

$$R\,|\ln(U',V';A',B')| = a$$

with the two solutions

$$(U',V';A',B') = e^{a/R}, \quad (U',V';A',B') = e^{-a/R}$$

one of them being >1 and one <1. It follows that we have exactly one solution corresponding to the fixed half-line of the axiom and V 5 is verified.

The verification of the next axioms requires some trigonometrical formulas for h-triangles, which should relate the h-characteristics of the sides of a triangle with the Euclidean values of the angle of the respective semicircles.

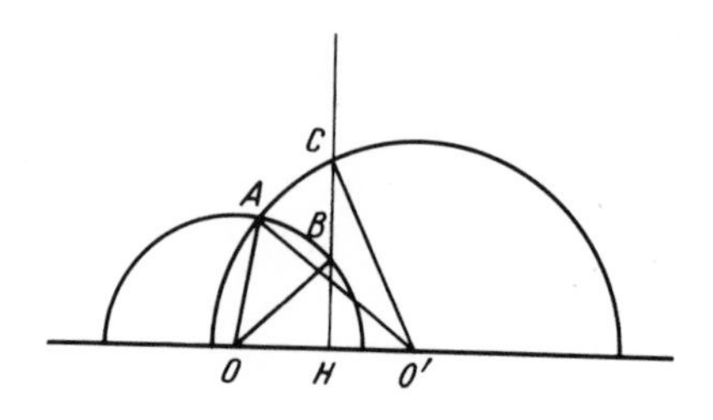

Fig. 90

We begin with an h-triangle ABC such that BC is represented by a half-line (Fig. 90), and we denote by $\hat{A},\hat{B},\hat{C}$ its Euclidean angles and by $a = h(BC)$, $b = h(AC)$, $c = h(AB)$ its sides. We also assume that, if O and O' are respectively the centres of the semicircles AB,AC and $H = OO' \cap BC$, then $H - B - C$ and $O - H - O'$. However, if O,H,O' have different positions, the proof is actually the same as the proof below.

Under our hypotheses, we have

$$a = R \mid \ln(H,V_\infty;B,C) \mid = R \ln \frac{HC}{HB}, \tag{7}$$

where, for the sake of simplicity, we denote the Euclidean lengths without l (i.e., $HC = l(HC)$, etc.).

It follows

$$e^{a/R} = \frac{HC}{HB},$$

whence

$$\operatorname{ch} \frac{a}{R} = \frac{HB^2 + HC^2}{2HB \cdot HC}. \tag{8}$$

By an examination of Fig. 90, we see that

$$HB^2 = OB^2 - OH^2 = OA^2 - OH^2$$

$$HC^2 = O'C^2 - O'H^2 = O'A^2 - O'H^2,$$

whence, by the generalized Pythagoras Theorem for the triangle OAO', we get

$$HB^2 + HC^2 = OA^2 + O'A^2 - (OH^2 + O'H^2)$$

$$= OO'^2 + 2OA \cdot O'A \cos \widehat{OAO'} - (OO'^2 - 2OH \cdot O'H)$$

$$= 2OA \cdot O'A \cos \widehat{OAO'} + 2OH \cdot O'H .$$

Now, formula (8) yields

$$\operatorname{ch} \frac{a}{R} = \frac{OA}{HB} \cdot \frac{O'A}{HC} \cos \widehat{OAO'} + \frac{OH}{HB} \cdot \frac{O'H}{HC} = \frac{OB}{HB} \cdot \frac{O'C}{HC} \cos \widehat{OAO'} +$$

$$+ \frac{OH}{HB} \cdot \frac{O'H}{HC} = \frac{1}{\sin BOH} \cdot \frac{1}{\sin \widehat{CO'H}} \cos \widehat{OAO'} +$$

$$+ \operatorname{ctg} \widehat{BOH} \cdot \operatorname{ctg} \widehat{COH},$$

which actually means

$$\operatorname{ch} \frac{a}{R} = \frac{\cos \hat{A} + \cos \hat{B} \cos \hat{C}}{\sin \hat{B} \sin \hat{C}} . \tag{9}$$

Now, if we want to calculate an arbitrary side, represented by some semicircle, of an arbitrary h-triangle, we can apply an inversion sending the semicircle to a half-line, such as to arrive at the situation of Fig. 90. According to the properties of inversions, the values of the h-sides and of the angles of the given h-triangle remain unchanged. This shows that, as a matter of fact, formula (9) holds for every h-triangle and every side of it.

Further, from formula (9) we may derive

$$\operatorname{sh} \frac{a}{R} = \left[\operatorname{ch}^2 \frac{a}{R} - 1\right]^{1/2} =$$

$$\frac{[\cos^2 \hat{A} + 2 \cos \hat{A} \cos \hat{B} \cos \hat{C} + \cos^2 \hat{B} \cos^2 \hat{C} - \sin^2 \hat{B} \sin^2 \hat{C}]^{1/2}}{\sin \hat{B} \sin \hat{C}} =$$

$$\frac{[Q]^{1/2}}{\sin \hat{B} \sin \hat{C}} ,$$

where one denotes

$$Q = \cos^2 \hat{A} + \cos^2 \hat{B} + \cos^2 \hat{C} + 2 \cos \hat{A} \cos \hat{B} \cos \hat{C} -1.$$

Because this quantity is symmetrical in $\hat{A},\hat{B},\hat{C}$, we get the following *theorem of the sinus*

$$\frac{\operatorname{sh} \frac{a}{R}}{\sin \hat{A}} = \frac{\operatorname{sh} \frac{b}{R}}{\sin \hat{B}} = \frac{\operatorname{sh} \frac{c}{R}}{\sin \hat{C}} = \frac{\sqrt{Q}}{\sin \hat{A} \sin \hat{B} \sin \hat{C}} . \tag{10}$$

Finally, using formulas (9) and (10), one gets by a straightforward verification the following *theorem of the cosinus*.

$$\operatorname{ch} \frac{a}{R} = \operatorname{ch} \frac{b}{R} \operatorname{ch} \frac{c}{R} - \operatorname{sh} \frac{b}{R} \operatorname{sh} \frac{c}{R} \cos \hat{A}. \tag{11}$$

We are now able to prove that the remaining congruence axioms hold in P.

For V 6, if we use the notation from the formulation of this axiom, we can proceed as follows: take a semicircle through A' with the centre on the ideal line and which forms in the half-plane s an angle equal to the angle between the semicircles AC and AB. On the semicircle constructed in s, take C' such that $h(A'C') = h(AC)$, which can be obtained as in the verification of Axiom V 5. Then, formula (11) applied to the h-triangle $A'B'C'$ yields $h(B'C') = h(BC)$, which shows that C' is the point needed in V 6. The uniqueness of this point follows by reductio ad absurdum and again using formula (11).

Finally, for V 7, we consider the h-configurations described in the axiom, using the same notation, and, by formula (11), we deduce that $(A,B,C) \equiv (A',B',C')$ implies $\hat{A} \equiv \hat{A}'$. Next, by repeated utilization of (11), for the h-triangles ACP and $A'C'P'$, we derive $CP \equiv C'P'$ as required.

Summing up, we get

1. THEOREM. P is a model of plane absolute geometry.

Let us note, as a consequence, that we can now introduce the measures of segments and of angles. Since $h(AB)$ has just the properties of the segment measure, we see that formula (5) defines the length of a segment in P and, similarly, the Euclidean measure of angles between semicircles is also the angle measure in P. This last property is very important for the utilization of P in the study of Lobachevski geometry. Actually, it means that P is a *conformal model* of the Lobachevski plane.

And now about parallels. It is easy to understand that in P the Euclidean parallel axiom does not hold. Indeed, if UV is an h-line and A an exterior h-point (Fig. 91), then the semicircles AU and AV with centres on the ideal line $\underline{a}$ are two parallels through A to UV and any semicircle like the dotted one of Fig. 91 is also parallel to UV.

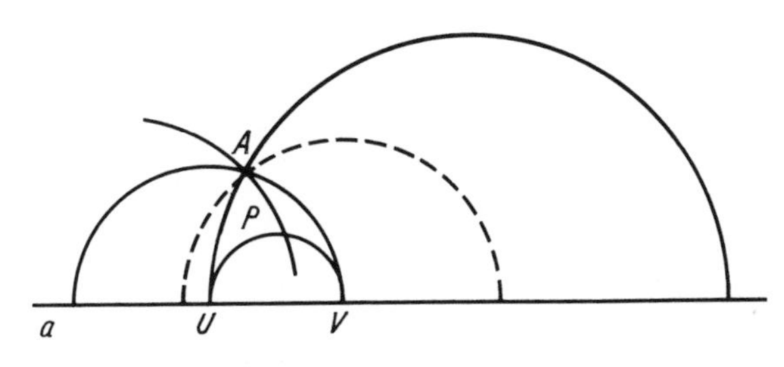

Fig. 91

Hence, in P, the negation of the Euclidean parallel axiom, i.e., the Lobachevski axiom, holds and we have

2. THEOREM. Plane Lobachevski geometry is consistent if plane Euclidean geometry is such.

We have already said that the Poincaré model may be used in developing Lobachevski geometry. For instance, formulas (9), (10) and (11), are just trigonometrical formulas in Lobachevski plane. It is worth making a few other remarks along this line.

Fig. 91 shows that there are two distinguished parallels through A to UV, namely, AU and AV. In fact, in Lobachevski geometry, the name of *parallels* is reserved exclusively for these two lines, while the other lines through A and not intersecting UV are called *divergent lines*. There are essential differences between the properties of parallel and divergent lines.

If we take the h-line $AP \perp UV$, then the angle between AP and AU or AV is called the *angle of parallelism*. It is possible to show that this angle depends on the h-distance from A to UV only. We can calculate it in the following way.

Without restricting the generality, we can suppose that we are dealing with the case of Fig. 92, where the perpendicular AP is represented by Euclidean half-line, $h(AP) = x$, $O - P - A$ (O is the centre of UV), $OU = 1$ and $OA = h$.

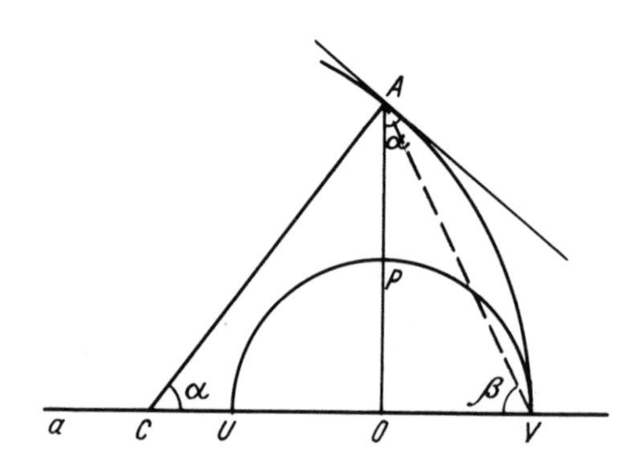

Fig. 92

Then, we have as for formula (7)

$$x = R \ln h = R \ln \operatorname{tg} \beta,$$

where β is the corresponding angle of Fig. 92.

But from the isosceles triangle ACV, where C is the centre of AV, we deduce

$$\beta = \frac{\pi}{2} - \frac{\alpha}{2},$$

where again α is defined in Fig. 92, whence, $x = \ln \operatorname{ctg} \alpha/2$, and

$$\alpha = 2 \operatorname{arc} \operatorname{tg} e^{-x/R}. \tag{12}$$

One usually denotes $\alpha = \Pi(x)$ and this is called the function of Lobachevski. It plays an important role in the geometry of Lobachevski.

Using again P, we can also get representations of some interesting curves. Let us define an *elliptic pencil* of lines as a family of concurrent h-lines, a *parabolic pencil* as a family of parallel (not divergent) lines and a *hyperbolic pencil* as a family of lines which are orthogonal to a given h-line. Then, the orthogonal trajectories of such pencils are respectively called a *circle*, a *horocycle*, and an *equidistant line*. One can see that such trajectories are represented by circles of P, or by parts of circles, whose centres are no more on the ideal line.

Finally, we would like to mention the availability of an interesting interpretation of the congruent transformations in Lobachevski plane. Namely, they are represented in P by products of inversions with poles on the ideal line and symmetries with respect to lines which are orthogonal to the ideal line.

Now, if we want to go to the three-dimensional case, the Poincaré model has the following generalization. Consider real Euclidean space and fix some plane π, called the *ideal plane*. Fix one of the half-spaces with origin π and call it the *upper half-space*. Define the *h-points* as the points of the upper half-space, the *h-lines* as semicircles and half-lines situated in the upper half-space and orthogonal to π, and the *h-planes* as hemispheres and half-planes of the upper half-space which are orthogonal to π.

Then, the verification of the spatial incidence axioms consists of a number of exercises in Euclidean geometry, which we leave for the reader to work out.

As for the other groups of axioms, they all refer to plane configurations and by means of a convenient inversion in space we can always assume that we work with an h-plane represented by a Euclidean half-plane. Then, everything is like in the model P of plane geometry.

Hence, in this manner, we get a model of three-dimensional absolute geometry, which does not satisfy the Euclidean parallel axiom, and this proves

3. THEOREM. The Euclidean parallel axiom is independent of the axioms of three-dimensional absolute geometry.

Of course, the same is true for the Lobachevski parallel axiom, i.e., this too is independent on the axioms of absolute geometry, because of the consistency of Euclidean geometry.

PROBLEMS

1. Show that the set $V_c(d)$ of the congruence vectors of a line d has a natural structure of an additive abelian group.
2. Show that, on every line d, the product of a symmetry with a translation is a symmetry.
3. Show, on the basis of Axioms I, III', V, only, that for any point M and any line d, if $M \bar{\in} d$ there is exactly one line d' such that $M \in d'$ and $d' \perp d$. If $M \in d$ then in each plane α containing d there is exactly one line d' with this property.
4. Show, based on I, III', V, that any congruent transformation between two planes sends two orthogonal lines to two orthogonal lines.
5. If α and α' are two planes and $d_1 \perp d_2$ are two lines of α, $d_1' \perp d_2'$ are two lines of α', then there are congruent transformations from α to α' which send d_1 to d_1' and d_2 to d_2'.

6. Show that an angle is a right angle iff it is congruent with its supplement. Show that any two right angles are congruent.

7. Start with Axioms I, III', V', and define the triangle congruence $ABC \equiv A'B'C'$ by the relations $AB \equiv A'B'$, $AC \equiv A'C'$, $BC \equiv B'C'$. Show that if for two triangles ABC, $A'B'C'$ one has $AB \equiv A'B'$, $AC \equiv A'C'$ and $\hat{A} \equiv \hat{A}'$ then the two triangles are congruent.

8. Show, on the same logical basis as in Problem 7 that, if for the triangles ABC and $A'B'C'$ one has $BC \equiv B'C'$, $\hat{B} \equiv \hat{B}'$, $\hat{C} \equiv \hat{C}'$, then the two triangles are congruent.

9. Use again I, III', V' and show on this basis that the proposition: "the corresponding angles of two congruent triangles are congruent" implies the fact that angle congruence is an equivalence relation.

10. Let a,b be two coplanar lines. Two points $A \in a$, $B \in b$ are called *homologous* if the transversal AB encloses, on the same side, equal interior angles with a and b. Show that when a and b have no intersection point every $A \in a$ has at most one corresponding homologous point on b.

11. Show that if two coplanar lines have a pair of homologuous points they have a symmetry axis. (N.B. One can actually show that any two coplanar lines without an intersection point have a unique symmetry axis. Of course, one has here to consider only incidence, order and congruence axioms.)

12. Let ABC be a triangle where $AC > BC$. Let D be the intersection point of the bisector of $\hat{C}$ with AB. Show that $AD > BD$.

13. A simple polygon $A_1A_2 \ldots A_n$ is called *regular* if $(A_1, A_2, \ldots, A_n) \equiv (A_2, A_3, \ldots, A_n, A_1)$. Show that the congruent transformations of the plane which fix a regular polygon form a finite subgroup of the congruence group of the plane of that polygon.

14. Prove that in a Euclidean space the perpendicular bisectors of the sides of a triangle are concurrent.

15. Prove that in a Euclidean space the altitudes of a triangle are concurrent.

16. Prove Theorem 1.2.13 (Desargues) in a Euclidean plane using triangle similarities.

17. Give a complete proof of Theorem 2.11 of this chapter (Pythagoras).

18. Let ABC be an arbitrary triangle and d a line not containing A, B,C and such that $d \cap AB = C'$, $d \cap BC = A'$, $d \cap CA = B'$. Show that the following relation (of Menelaus) holds:

$$AB' \cdot BC' \cdot CA' = CB' \circ BA' \cdot AC'.$$

19. Let ABC be an arbitrary triangle, and $A' \in BC$, $B' \in CA$, $C' \in AB$ three points such that AA', BB', CC' are concurrent lines. Show that the following relation (of Ceva) holds:

$$AB' \circ BC' \cdot CA' = CB' \cdot BA' \cdot AC'.$$

20. Show that the field $k(R,\text{id.})$ of the formal power series with real coefficients is an Euclidean field.

21. Define a geometry consisting of an affine space and a quadratic form such that non-zero vectors with zero length exist. Show that if we take all these vectors starting from the same origin O we get a cone (i.e., a set of points C containing O and such that if $X \in C$ the line OX is contained in C).

22. Suppose that we admit Axioms I, III, V and the axiom of Archimedes of Section 2. Then, show that, if $\alpha > \varepsilon$ are two arbitrary acute angles, there is a natural number n such that $2^n \varepsilon > \alpha$.

23. Prove that a congruent transformation of the space of absolute geometry is continuous.

24. Prove in absolute geometry that if a line d passes through an interior point P of a circle C, then d intersects C in exactly two points.

25. Show that the perpendicular bisector of the basis of a Saccheri quandrangle is a symmetry axis of the quadrangle. (In absolute geometry).

26. Prove that the perpendicular bisector of a side of a triangle ABC is orthogonal to the line joining the middles of the two other sides. (In absolute geometry).

27. Prove Proposition 3.16 of this chapter.

28. Prove that in a Euclidean space perpendicularity of lines is a binary relation $\perp$ on the set of lines, which satisfies the following four conditions:

    (1) $d \perp d'$ implies $d' \perp d$;

    (2) $d_1 \perp d_2$ and $d_1' \parallel d_1$ imply $d_1' \perp d_2$;

    (3) the locus of the lines which contain a given point O and are $\perp$ to a given line d is a plane α;

    (4) $d \perp d$ never holds.

    (Recall that $d_1 \perp d_2$ in space if the angle between a vector of d_1 and a vector of d_2 is a right angle.)

29. Suppose we have an affine space endowed with a binary relation $\perp$ for lines, satisfying (1), (2), (3), (4), of Problem 28. Define: $AB \equiv CD$ if taking some point O and constructing $\overrightarrow{OM} \sim \overrightarrow{AB}$, $\overrightarrow{ON} \sim \overrightarrow{CD}$ one has $OP \perp MN$, where P is the middle of MN (in the affine sense). Show that this congruence relation does not depend on the choice of the point O.

30. Consider an ordered affine space with a relation $\perp$ like in Problem 29, and take an affine frame $\{O, \bar{e}_i\}$ $(i = 1,2,3)$ such that its axes are pairwise orthogonal and the segments defining the $\bar{e}_i$ are congruent in the sense of Problem 29. One can show [16], that with respect to such a frame, the equation of the plane through O and containing the perpendiculars to the vector $\bar{a}(a_i)$ $(i = 1,2,$ $3)$ is

$$x^1 a_1 + x^2 a_2 + x^3 a_3 = 0.$$

Supposing that we already know this result, show that the defined congruence gives the space a structure of a Euclidean space.

31. Prove, using the Poincaré model, that two lines of the Lobachevski plane are divergent iff they have a common perpendicular line.

HINTS FOR SOLVING THE PROBLEMS

Chapter 1

1. Take the pairs of the points as lines.
2. Pairs of points as lines and triples as planes.
5. Take the affine plane over the integers modulo 3. It is impossible to get models with less than 9 points.
6. As for Problem 5.
7. In the case of the usual incidence axioms the scalar field is given by the integers modulo 2 and, in the case of the strong axioms, by the integers modulo 3.
8. The lines through a point are the lines joining the point to points of another line and the parallel to the last.
9. (c), (d) Use an affine frame.
10. Use the definition of the product of two matrices and show that each entry of MM' is equal to the corresponding entry of $nI_{n^2} + J_{n^2}$.
12. The axioms are those contained in Theorem 2.10. They are verified in the model by classical theorems on linear homogeneous equations.

15. The necessary axioms are verified by choosing each time a conveniently large n.

16. By a reductio ad absurdum. A contradiction will be derived if one considers for the elements of the configuration the smallest n such that the corresponding element is in the M_n of the free plane.

17. The free plane of this problem must be infinite (which follows from the foregoing problems) and this gives the possibility to construct 10 points and 10 lines like in the Desargues configuration. If one had the Desargues theorem, this configuration is bounded, which contradicts Problem 16.

(For Problems 14-17 see, e.g., R. Hartshorne, *Foundations of Projective Geometry*, Benjamin Inc., New York, 1967).

20. To get the middles, we must consider parallels through the vertices to the opposite sides. Thereby we get a new triangle and we can apply the converse of the Desargues Theorem. The second assertion follows using again, in a convenient manner, the Desargues Theorem.

24. Use the classical decomposition of a vector by projections on two axes.

25. Use parametric equations of the planes.

27. Substract the first line of the mentioned matrix from all the other lines. Then, if the condition on the rank does not hold the lines of the new matrix are linearly dependent, which yields the desired result.

29. Take an affine frame with the origin A, and the basis $(\bar{e}_1, \bar{e}_2)$, where $\bar{e}_1 = \overrightarrow{AB}$ and $\bar{e}_2 \parallel l$.

30. Use the computations of Problem 29.

31. Use again a conveniently chosen affine frame.

(For the problems 29-31 see also E. Artin [1] and the previously mentioned book by R. Hartshorne).

32. The first assertion is by a straightforward verification. The second requires constructing a concrete example.

33. In the plane, for instance, we count the transformations by counting the triangles, because they may be defined by the image of a fixed triangle. In the space, we use tetrahedra.

34. For the second part, use coordinates with corresponding origins and bases of vectors.

35. Recall that M is a fixed point of a transformation α if $\alpha(M) = M$ A similar definition is for lines. (However, not every point of a fixed line is a fixed point!)

36. Use a coordinate system.

Chapter 2

1, 2 and 3. Choose a point not on the line of the given points, join it with the given points and, on one of these lines, take a third point (the line on which we take this point must be conveniently chosen). Next, use several times the Pasch Axiom. The proof of Proposition 2.§1.4 will be taken as a model.

4. The proof is by induction on n and is based on the results of Problems 2 and 3 above.

5. Take another point E and put in class I every point A such that $\overline{A - O - E}$ and in class II every point B such that $B - O - E$.

6. Take a line d and a point $O \in d$. Denote the two half-lines, defined by O, by I and II. Put $A < B$ if either $A \in$ I, $B \in$ II or A, $B \in$ I and $A - B - O$ or $A, B \in$ II and $O - A - B$. Show that this yields the desired ordering relation.

8. If, for instance, v and w are on the same half-plane with respect to u, we have necessarily v in the interior of $\widehat{u,w}$ (or w in $\widehat{u,v}$) and the desired line is obtained by joining a point of u with a point of w.

9, 10, and 11. Analyze the half-planes defined by the respective lines. For the last part of Problem 11, use induction on n.

13. DE cannot intersect any other side of the polygon because D and E belong to different sides. The results follow then by analyzing the various appearing half-planes.

15. Start with a triangle and use the strong incidence axioms to construct a simple and non-convex quadrangle.

18. Glue up along a face two polyhedra like in Fig. 51, then three, four,..., such polyhedra.

21. The corresponding scalar field is of characteristic zero.

23. Use the same frame like for Problem 29 of Chapter 1 and the definition of the ordering relation for scalars.

24. Take the intersection of all convex sets containing M.

26. Obtain the necessary tetrahedra starting with a vertex of the convex envelope of M.

27. Show that a tetrahedron can be inscribed in a parallelepiped and conversely.

31. Take the triangles ABC, $A'B'C'$ such that AA', BB', CC' be distinct lines, concurrent at O, and $AB \parallel A'B'$, $AC \parallel A'C'$. Suppose OB and AC are not parallel. Let AML be the parallel through A to OB and with $L \in A'C'$, $M \in OC$. Let be $N = LB' \cap AB$. Use the Pappus Theorem for the following pairs of triples: (O,A,O'), (B',N,L); (A,B,N), (O,M,C); (O,M,C'), (L,B',N). It will follow $BC \parallel B'C'$ as required. If $OB \parallel AC$, one changes OBB' by ODD' with $D \in AB$, $D' \in A'B'$, and the same conclusion follows. A similar idea helps proving the result for the case when $AA' \parallel BB' \parallel CC'$.

33. Project on the axes of an affine frame and use the linear Bolzano-Weierstrass theorem 2.4.5.

35. Take $A - D - C$, $B - E - C$ and construct the desired neighborhoods on different parts of the line DE.

38. For example, the field of the real numbers of the form $a + b\sqrt{2}$ with a,b-rationals defines such a geometry.

Chapter 3

3. If $M \bar{\in} d$, take as M' the image of M by the symmetry with respect to d and show that d' is just MM'. If $M \in d$, take $PM \equiv MQ$, $P - M - Q$ on d and construct the axis of the symmetry which sends P to Q.

4. This is because a right-triangle is sent to a right-triangle.

5. Take $O = d_1 \cap d_2$, $O' = d_1' \cap d_2'$, $A \in d_1$, $B \in d_2$, $A' \in d_1'$, $B' \in d_2'$ such that $OA \equiv O'A'$, $OB \equiv O'B'$, and take the congruences which send A to A' and B to B'.

6. Let the angle be $\widehat{hOk}$ and Ok' the prolongation of Ok. Use equal segments on the half-lines h,k,k' to prove the first result, and use Problem 5 for the second result.

7. Suppose $BC \not\equiv B'C'$ and take C'' such that $\overline{C - B - C''}$ and $BC'' \equiv B'C'$. Use axiom V' 5 to establish $\widehat{BAC''} \equiv \widehat{B'A'C'}$, which together with $\hat{A} \equiv \hat{A}'$ contradicts the uniqueness assertion of axiom V' 4.

8. The proof is by the same method as in Problem 7 above.

10. Use the External Angle Theorem.

11. The desired symmetry axis is the symmetry axis of the given pair of homologous points.

12. Take on CB the point A' such that $CA' \equiv CA$. It follows $AD \equiv A'D$, $\widehat{CAD} \equiv \widehat{CA'D}$. On the other side, $\widehat{DBA} > \widehat{CAD}$ by the External Angle Theorem. Hence, $A'D > BD$, because their opposite angles in the triangle DBA' are in this relation, and we get the required inequality.

13. The result follows from the fact that such a transformation induces a permutation of the vertices of the polygon.

15. The altitudes of the given triangle are the perpendicular bisectors of the triangle obtained if by each vertex of the initial triangle we draw a parallel to the opposite side. Hence, the result follows by Problem 14.

18. Consider the perpendiculars from A, B, and C to d and use the similarities of the obtained right triangles.

19. Use the Menelaus relation for the triangle $AC'C$ and the line BB' and, next, for the triangle $BC'C$ and the line AA'.

21. It suffices to take a non-degenerate quadratic form which is neither positive nor negative definite.

22. Let be $\alpha = \widehat{hOk}$, $A \in h$, $AB \perp k$, $B \in k$, C_1 the intersection point of AB with the bisector of $\widehat{hOk}$, C_2 the intersection point of C_1B with the bisector of $\widehat{BOC_1}$, etc. In this manner, and using the results of Problem 12, we get a point C_n such that $\widehat{BOC_n} < \varepsilon$, which proves the result.

23. Use triangular neighborhoods and the definition of a congruence.

24. Consider a half-line s of d with the origin P and define the function $\varphi(M) = OM$, the distance from the centre O of C to $M \in s$. Then $\varphi(P) < r$, where r is the radius of C. Show that there is some $M_0 \in s$ with $\varphi(M_0) > r$ and that φ is continuous. This implies the existence of an intersection point $s \cap C$. Finally, use similarly the second half-line of d with origin P.

26. Use the same construction as in the proof of Proposition 3.3.14 to derive a convenient Saccheri quadrangle and, next, use the result of Problem 25, Chapter 3.

28. Take an orthonormal frame and two vectors $\bar{v}(\xi^i)$, $\bar{w}(\eta^i)$. Show, using the Pythagoras Theorem, that they are orthogonal iff $\sum_{i=1}^{3} \xi^i\eta^i = 0$. Now, we easily get the stated properties.

29. Use the Desargues Theorem.

30. Fix an orthogonal frame and consider two points $M(\xi^i)$, $N(\eta^i)$. Then, the coordinates of $\overrightarrow{MN}$ are $\eta^i - \xi^i$ and those of the midpoint

P of MN are $(\eta^i + \xi^i)/2$. $OM \equiv ON$ iff OP belongs to the plane through O and perpendicular to MN, which, in view of the known equation of that plane gives

$$\Sigma(\xi^i)^2 = \Sigma(\eta^i)^2.$$

Hence, congruence of segments is characterized by means of the well-known scalar characteristic, and the Euclidean structure of space follows.

31. Using an inversion, we can always consider the case when one of the two lines mentioned is represented in the model by a half-line perpendicular to the absolute line. In this case, the result follows by elementary geometry.

REFERENCES

1. E. Artin, *Geometric Algebra*, Intersc. Publ. Inc., New York, 1957.
2. R. Artzy, *Linear Geometry*, Addison-Wesley Publ. Comp., Reading Mass., 1974.
3. K. Borsuk, W. Szmielew, *Foundations of Geometry*, North-Holland Publ. Comp., Amsterdam, 1960.
4. N. Bourbaki, *General Topology*, Addison-Wesley Publ. Comp., Reading, Mass., 1966.
5. H.S.M. Coxeter, *Non-Euclidean Geometry*, Univ. of Toronto Press, Toronto, 1965.
6. N.V. Efimov, *Vysšaia Geometria*, Gos. Izd. Fiz. Mat. Literatury, Moskwa, 1961.
7. S. Feferman, *The Number Systems*, Addison-Wesley Publ. Comp., Reading, Mass., 1963.
8. G.M. Fikhtengoltz, *The Fundamentals of Mathematical Analysis*, Oxford, Pergamon Press, 1965.
9. D. Hilbert, *Foundations of Geometry*, La Salle, Ill., Open Court, 1971.
10. B. Kerékjártó, *Les Fondements de la Géométrie*, I. Akademiai Kiadó, Budapest, 1955.

11. S. Lang, *Algebra*, Addison-Wesley Publ. Comp., Reading, Mass., 1965.

12. P.K. Rashewski, *Osnovania Geometrii Hilberta i ih mesto v istoriceskom razvitii voprosa.* Introduction to the Russian translation of D. Hilbert's *Grundlagen der Geometrie*, Gostehizd., Moskwa, 1948.

13. B.A. Rozenfeld, *Neevklidovy Prostranstva*, Nauka, Moskwa, 1969.

14. E. Snapper, R.J. Troyer, *Metric Affine Geometry*, Academic Press, New York, 1971.

15. E. Steinitz, H. Rademacher, *Vorlesungen uber die Theorie der Polyeder*, Springer-Verlag, Berlin, 1934.

16. Gh. Vranceanu, C. Teleman, *Geometrie euclidiana, Geometrii neeuclidiene, teoria relativitatii,* Ed. stiintifica, Bucuresti, 1965.

17. Gh. Vranceanu, Th. Hangan, C. Teleman, *Geometrie elementara din punct de vedere modern,* Ed. Tehnica, Bucuresti, 1967.

18. H. Weyl, *Raum-Zeit-Materie*, Springer-Verlag, Berlin, 1923.

INDEX